CELL LOCOMOTION *IN VITRO*

CELL LOCOMOTION *IN VITRO*

Techniques and Observations

C. A. Middleton and J. A. Sharp

UNIVERSITY OF CALIFORNIA PRESS
Berkeley and Los Angeles

University of California Press,
Berkeley and Los Angeles, California

Library of Congress Cataloging in Publication Data
Middleton, C.A..

Cell locomotion in vitro.
Includes index.
1. Cells–Motility. 2. Cell culture.
[DNLM: 1. Cell movement. 2. Cells, cultured.
3. Cells–Physiology. QH 647 M628C]
QH647.M52 1984 574.871'64 83-18215

ISBN 0-520-05209-9

Printed and bound in Great Britain

CONTENTS

PREFACE

It is ten years since the first symposium on cell locomotion was held (*Locomotion of Tissue Cells*, Ciba Foundation Symposium 14, 1972). That meeting was chaired by Michael Abercrombie, and in his introductory remarks he commented on the extent to which the importance of cell locomotion, apart from that seen in leucocytes, had been underestimated. Much has been done to correct that neglect during the succeeding decade, and we have learned more about the underlying mechanisms of cell locomotion and about the factors which may influence it. Abercrombie was himself a major contributor to this field of research (as a glance at the lists of references in this book will confirm), and his ideas inspired the work of many other investigators.

As in all branches of science, progress in the study of cell locomotion has depended on the availability of appropriate experimental techniques. Of these, tissue culture has made the greatest contribution, in conjunction with a variety of procedures using either the light or the electron microscope. We have, therefore, attempted, in chapters 2 and 3, to provide explanations of the techniques which have been particularly fruitful, but only in sufficient detail to permit the reader to grasp their essentials; this book is not a laboratory manual.

In the remaining chapters our aim has been to present an outline of the existing state of knowledge about the locomotion of cells in culture and the mechanisms which may subserve it. This is intended for undergraduates, or those embarking on a postgraduate course, or anyone in need of a succinct account of this important area of biological research.

ACKNOWLEDGEMENTS

We thank Dr Jeremy Hyams for reading, and advising us on, part of the manuscript; remaining errors or misconceptions are entirely our own. We thank all those who have made illustrative material available to us; they are named individually in the legends to the figures. We also thank Mrs C.A. Peters and Mrs H.M. Sharp for the skill and care with which they prepared the typescript.

1 INTRODUCTION

Locomotion is defined by the *Oxford English Dictionary* as 'the action or power of moving from place to place'. This book is an attempt to outline our knowledge of the mechanisms which enable living cells to move from place to place within the body of an animal, and to explain some of the experimental techniques which have contributed to this knowledge.

Cell locomotion is, of course, only one aspect of the more general phenomenon of cellular motility which has been studied in systems as diverse as bacterial flagella, plant cytoplasm and striated muscle. We have chosen to concentrate here on the processes associated with the movement of cells in tissue culture. Although we shall make use of data derived from other experimental systems, we feel that, at present, the study of living cells moving in tissue culture offers the most fruitful approach to understanding the mechanisms underlying, and the influences which control, cell locomotion *in vivo*.

Apart from the basic desire to satisfy their curiosity about a particular natural phenomenon, there is another, perhaps more cogent, reason why biologists wish to learn as much as possible about cell locomotion. The active migration of cells is of basic importance in the establishment of the structure of the body during embryonic development; after birth, it is equally vital in the healing of wounds and the protection of the body against infection by bacteria and viruses; and it is one of the most sinister features of the abnormal behaviour of cancer cells. Hence a better understanding of cell locomotion, clarifying both the intrinsic mechanisms within the cell and the factors which normally control the movement, is likely to contribute significantly to research in embryology and immunology, and to the study of the behaviour of cancer cells, three areas of biology in which fundamental problems remain to be solved.

Cell Locomotion in Embryology

Embryology furnishes numerous striking examples of shifts of tissues relative to each other, moving either as compact masses or as groups of individual cells. Indeed, the processes molding the early

> embryo after cleavage are predominantly in the nature of translocations rather than growth. Practically the whole germ is on the move. Later, after the basic form has become fixed, mobility is restricted to certain cell types which move within the now consolidated frame. The neural crest, for instance, spreads into the interstices of the embryonic body, laying down different cell types at different stations: ganglion cells along the vertebral column, sheath cells along the nerve fibers, pigment cells along predetermined lines in the integument and its derivatives, and, at least in Amphibians, cartilage for certain elements of the head skeleton. There is circumstantial evidence that these various cell types are already different in character when they leave their common sites of origin. What, then, guides each to its proper final destination? (Weiss, 1947)

Since these words were written, the movement of groups of cells from place to place within the embryo has been the subject of continuing research by experimental embryologists. Confirmation of the importance of the particular example cited by Weiss, the migration of neural crest cells, has recently been provided by an elegant technique applied at the Institut d'Embryologie at Nogent-sur-Marne by Le Douarin and Le Lièvre. Their method relies upon the fact that there are easily recognisable differences between the interphase nuclei in two related species of birds, the Japanese quail (*Coturnix coturnix japonica*) and the chick (*Gallus gallus*) (Le Lièvre and Le Douarin, 1975). By excising a piece of the neural tube and associated neural crest from a chick embryo and transplanting the corresponding portion of a quail embryo of the same age, it is possible to follow the subsequent migration of the quail neural crest cells within the chick embryo and to define precisely the eventual distribution and differentiation of these cells. Such experiments have established that neural crest cells spread out to form a variety of tissues, including, for example, the visceral skeleton, the dermis of the face and ventrolateral side of the neck, the walls of arteries, and the connective tissues in the thymus, thyroid and parathyroid glands (Figure 1.1). Although they provide conclusive proof of the extent and the importance of the contribution made by the neural crest, these experiments do not, of course, throw any light on the mechanisms which operate within the cells during their migration, nor do they offer any answer to the fundamental question posed by Weiss – what guides each to its proper final destination?

Figure 1.1: Diagram showing the migration of neural crest cells in the head and branchial arch regions of a chick embryo. Cells from the prosencephalon migrate to the frontal region (Pro.N.C.), those from the mesencephalon are found in the facial area (Mes.N.C.), while rhombencephalic neural crest cells migrate into the branchial arches (Rho.N.C.). (Redrawn from Le Douarin, 1979.)

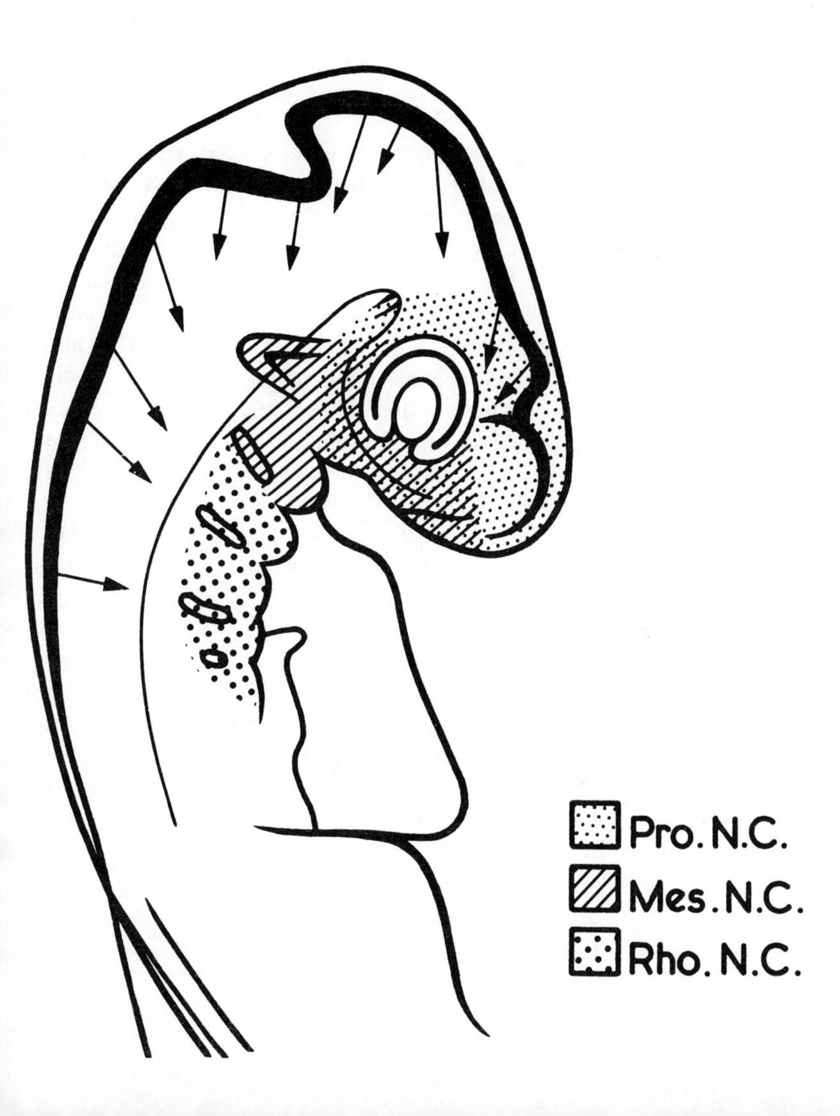

Cell Locomotion in Immunity

In vertebrate animals, cells of two different kinds are involved in the initiation of an immune response. Lymphocytes are essential, since they appear to be the only cells capable of a specific immunological response to an antigen. Macrophages, while perhaps not strictly essential, certainly make a very important contribution by promoting the efficient functioning of the lymphocytes.

Judging by their behaviour *in vitro*, lymphocytes and macrophages are amongst the most rapidly motile cells in the body; their locomotory behaviour has been widely studied by means of time-lapse cinemicrography of cultures of lymphoid and other tissues. R.J.V. Pulvertaft produced one of the earliest systematic studies of the behaviour of lymphoid cells in culture, and his films led him to comment (1959):

> When you see the lymphocyte moving in a time lapse or accelerated film, you may be reminded of Graves' poem on the white butterfly:
> 'The butterfly, the cabbage white,
> (His honest idiocy of flight)
> Will never now, it is too late,
> Master the art of flying straight.'

Figure 1.2: Changes in the position and shape of a single living lymphocyte recorded at intervals of 0.5 minutes. (Redrawn from Nunn *et al.*, 1970.)

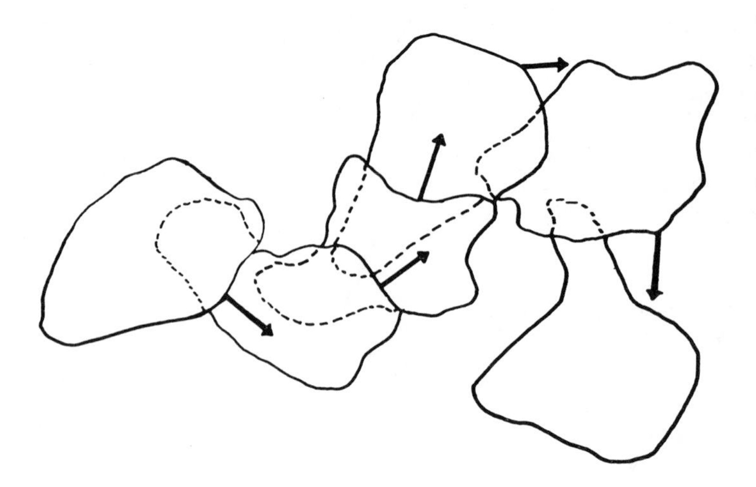

Figure 1.2 illustrates the movement of a lymphocyte during a period of 2.5 minutes, and shows the typical 'hand-mirror' shape assumed by the cell during locomotion, with a broad veil of cytoplasm at its advancing edge and a narrow 'handle' of cytoplasm trailing behind. We should perhaps picture the lymphoid tissues and organs as containing a seething mass of wriggling lymphocytes and macrophages. Pulvertaft speculated about the possible implications of the locomotion of the lymphocyte:

> From the enormous literature relating to this cell I shall extract at first two established facts only: first, that lymphocytes are not uniformly distributed in the tissues, but characteristically are aggregated in foci; secondly, that they are motile. Motility in cells must imply a function; we are in no doubt what that function is in the case of spermatozoa, polymorphs and monocytes. When motility ceases, we must assume that the objective has been attained, as journeys end in lovers meeting. The cliché 'round cell infiltration' establishes the fact that the wanderings of the lymphocyte are not invariably aimless, but that under certain conditions they clearly reach their destination and cease to move. (Pulvertaft, 1959)

During the last 20 years, immunology has become one of the most rapidly growing subjects in biology, and hundreds of scientific papers have been published bearing on the behaviour and functions of lymphocytes and macrophages. Pulvertaft's guesses about the probable importance of the powers of locomotion of these cells, and the ability of the lymphocyte to migrate to a specific destination, have been vindicated. For example, it has been shown that T-lymphocytes are able to congregate in the innermost part of the cortex of a lymph node by migrating through the walls of specialised postcapillary venules, and that cytotoxic T-lymphocytes, by virtue of their motility, are enabled to seek out 'foreign' cells and kill them by making contact with their surface. If lymphocytes were not capable of controlled locomotion, immune responses, especially of the cell mediated variety, could not occur.

Cell Locomotion in Cancer

The two most sinister attributes of malignant cancer cells are, first, their capacity for uncontrolled mitotic division, and second, their ability to spread from the original site of the tumour to distant parts of the body. The latter process, known as metastasis, is frequently the ultimate cause of the death of the patient due to the passage of cancer cells to a vital organ such as the brain or the liver.

Metastasis is a complex phenomenon; at the risk of oversimplification it may be said to involve three stages:

1. The detachment of cells from the primary tumour, and their movement into adjacent blood vessels or lymphatics, or into a body cavity.
2. The passive carriage of the cells via the vascular or the lymphatic system, or their dispersal within a body cavity, to a distant site.
3. Their emergence from a blood vessel or lymphatic and invasion of the contiguous tissue or organ.

It is generally agreed that the intrinsic motility of cancer cells plays an important part, especially in the first and third stages of metastasis. Locomotion by cancer cells has repeatedly been observed in tissue culture, but it is obviously more difficult to demonstrate unequivocally that they are in fact capable of locomotion *in vivo*. Circumstantial evidence of such behaviour is provided by histological sections in which cells appear to be migrating away from the edge of a tumour (Figure 1.3), though it may be argued that this is merely the result of the passive extrusion of cells caused by the pressure exerted by the expansion of the tumour. However, the fact that some tumours which grow slowly may invade and metastasise more extensively than others which grow rapidly suggests strongly that active locomotion is indeed of primary importance in metastasis (Willis, 1967).

Direct evidence that tumour cells are capable of escaping from capillaries *in vivo* by migrating through their walls was produced by Wood (1958). He inserted a transparent chamber into the ear of a rabbit; when capillaries had grown into the cavity within the chamber, he injected a suspension of cancer cells into the artery supplying the ear, and was able to produce a cinematographic record of the adhesion of the cells to the capillary endothelium and their migration, within a few hours, through the capillary wall into the extravascular space.

Advances in our knowledge of the underlying mechanisms of cell locomotion, and of the factors which initiate and arrest it or which influence the direction of the movement, must surely contribute fundamentally to the solution of outstanding problems in the fields of embryology and immunology, and could conceivably be of profound importance in the control of cancer.

Figure 1.3: Pale staining cancer cells invading and disrupting dark staining smooth muscle fibres in a carcinoma of the prostate. (Courtesy of Dr C.K. Anderson.)

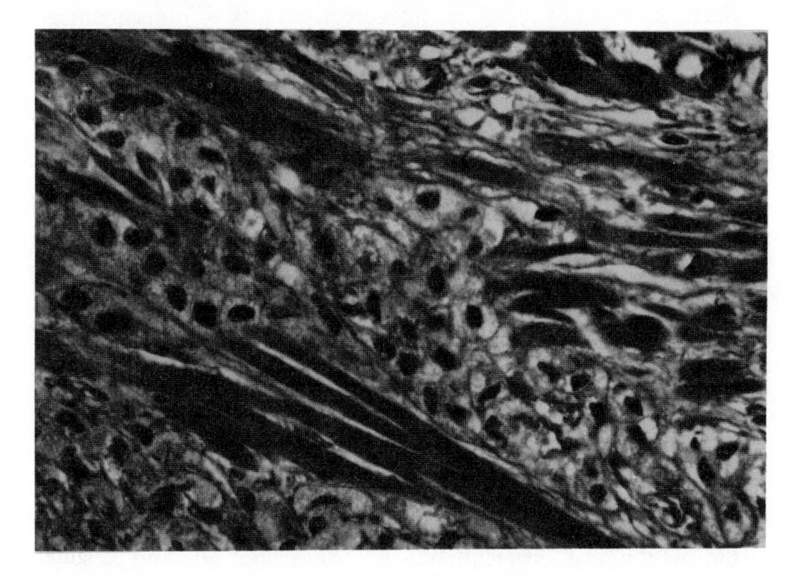

References

N.M. Le Douarin (1979) 'Dependence of Myeloid and Lymphoid Organ Development on Stem-Cell Seeding: Investigations on Mechanisms in Cell-Marker Analysis', in J.D. Ebert and T.S. Okada (eds), *Mechanisms of Cell Change* (John Wiley and Sons, New York), pp. 293-326

C.S. Le Lièvre and N.M. Le Douarin (1975) 'Mesenchymal Derivatives of the Neural Crest: Analysis of Chimaeric Quail and Chick Embryos', *J. Embryol. exp. Morph.*, vol. 34, p. 125

J.F. Nunn, J.A. Sharp and K.L. Kimball (1970) 'Reversible Effect of an Inhalational Anaesthetic on Lymphocyte Motility', *Nature*, vol. 226, p. 85

R.J.V. Pulvertaft (1959) 'Cellular Associations in Normal and Abnormal Lymphocytes', *Proc. roy. Soc. Med.*, vol. 52, p. 315

P. Weiss (1947) 'The Problem of Specificity in Growth and Development', *Yale Journal of Biology and Medicine*, vol. 19, p. 235

R.A. Willis (1967) *Pathology of Tumours*, 4th edn. (Butterworths, London), p. 146

S. Wood, Jr. (1958) 'Pathogenesis of Metastasis Formation Observed in Vivo in the Rabbit Ear Chamber', *Arch. Pathol.*, vol. 66, p. 550

2 THE CULTIVATION OF CELLS *IN VITRO*

Tissue culture techniques are widely used in the study of cell locomotion, since they make it possible to observe living cells and record their behaviour. It was the need to make the growth of embryonic nerve fibres visible which led Ross G. Harrison to devise, in 1907, the first successful method for the cultivation of living tissue outside the body (i.e. '*in vitro*' as opposed to '*in vivo*'). This he did by excising pieces of nervous tissue from frog embryos and transferring (or 'explanting') them to a drop of clotted lymph on a coverslip sealed on to a cavity slide (the 'hanging drop' method), and watching the elongation of the developing axons as they emerged from the explanted spinal cord into the surrounding clot.

Harrison made no further use of the method he had painstakingly contrived, but he had some idea of its potential importance:

> This method, which obviously has many possibilities in the study of the growth and differentiation of tissues, has two very distinct advantages over the methods of investigation usually employed. It not only enables one to study the behaviour of cells and tissues in an unorganized medium free from the influences that surround them in the body of the organism, but it also renders it possible to keep them under direct continuous observation, so that all such developmental processes as involve movement and change of form may be seen directly instead of having to be inferred from series of preserved specimens taken at different stages. (R.G. Harrison, 1910)

Harrison's main interest was in embryology, and he naturally assessed the value of tissue culture from an embryologist's point of view; he could hardly have foreseen the extent to which it would become a basic tool in so many different biological disciplines, such as biochemistry, physiology, microbiology, pathology, immunology, genetics and cell biology.

After Harrison had shown that tissue culture *in vitro* was a practical possibility, capable of making a significant contribution to knowledge, the responsibility for the next phase in its development passed to Montrose T. Burrows and Alexis Carrel. Like Harrison, these two scientists were working in America, and Burrows visited Harrison to learn his technique, with the intention of applying it to the study of tissues from warm-blooded animals; this soon resulted in the successful growth of chick embryo tissue, using, instead of frog lymph, clotted chicken plasma, which was firmer and more easily obtainable. Burrows then collaborated with Carrel, using clotted plasma in Harrison's hanging drop method, on the cultivation of mammalian tissues, and they discovered that growth *in vitro* could be stimulated and prolonged by transferring portions of an existing culture to fresh plasma on a clean coverslip – a procedure known as 'subculturing'. Carrel was also responsible for the introduction of an extremely rich nutrient prepared by extracting minced chick embryo tissues with a simple saline solution ('chick embryo extract'); when used in combination with chick plasma this stimulated growth and cell division in the culture. Alexis Carrel was a surgeon, and his experience of aseptic procedures in the operating theatre led him to adopt a similar approach in the tissue culture laboratory, including the wearing of hooded gowns. He stipulated that 'The culture must be made in a warm, humid operating room with the same care and rapidity as a delicate surgical operation . . . the perfect teamwork of well-trained assistants is necessary' (Carrel and Burrows, 1911). His elaborate methods may have discouraged others from embarking on the application of tissue culture techniques by fostering the impression that they were complicated and difficult.

It was certainly very time-consuming to establish and maintain large numbers of hanging drop cultures, and, in spite of careful asepsis, cultures were lost because they became infected. These difficulties were significantly reduced when, in 1923, Carrel devised a flask which allowed cultures to be manipulated more easily and with less chance of contamination, as well as providing a larger volume of nutritive medium (the Carrel flask, Figure 2.1).

An important contribution to tissue culture technique came from W.H. and M.R. Lewis, who investigated the suitability of saline solutions, already in use by physiologists for the perfusion of excised organs, as media for the cultivation of tissues *in vitro*:

> Having found that tissues from chick embryos grow readily in artificial media, nutrient agar and bouillon, the next step was the attempt to cultivate such tissues in media all the constituents of which are known. In collaboration with Dr. Arthur Koelker we were able to obtain excellent growth in combinations of amido-acids and polypeptids of known constitution, in Locke's solution. Locke's solution with maltose and dextrose was then tried with such good results that

> Locke's solution [alone] was attempted with the very startling discovery that these embryonic tissues grow readily in the Locke's solution and various combinations of NaCl, $CaCl_2$, KCl and $NaHCO_3$ . . .
>
> It is to be hoped that an artificial medium will be found as satisfactory as the plasma, for the advantages are obvious if one can work with a known medium in the investigation of the many problems, which suggest themselves. (M.R. Lewis and W.H. Lewis, 1911)

Thus the first steps were taken in a quest which eventually led to a clearer understanding of the nutritional requirements of cells *in vitro* and the production of 'synthetic' media, whose composition is precisely known.

Figure 2.1: The Carrel flask. Several explants are positioned on the floor of the sterile glass flask, and immersed in a mixture of chick embryo extract and cockerel plasma

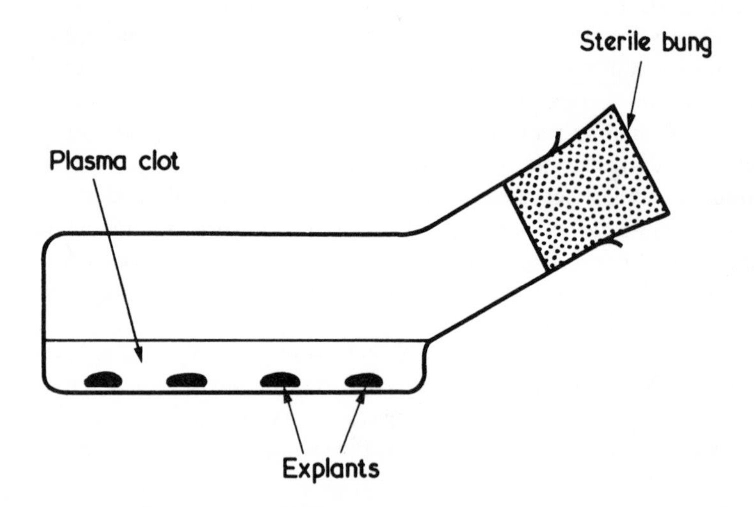

The hanging drop preparation, together with the Carrel flask, continued to be the most widely used methods in tissue culture until 1933, when G.O. Gey introduced the 'roller tube', in which explants of tissue were attached to the inner surface of a test tube with clotted plasma; a few millilitres of liquid medium were then added to the tube, which was closed with a bung and placed horizontally in a rotating drum inside an incubator. As the tube was rotated, the culture was alternately immersed in liquid medium and exposed to the air within the tube. Thus uniform distribution of the components of the medium was ensured, and long-term maintenance of cultures was made easier. It is to G.O. Gey that we also owe the introduction of collagen gel as a substratum for the growth of cells as a monolayer, and the establishment, in the early 1950s, of the first cell line (i.e. a population of cells with the potential to be subcultured indefinitely *in vitro*), derived from a case of cancer of the cervix, and known as HeLa cells (a contraction of Henrietta Lacks, the name of the patient from whom the original biopsy was obtained).

It was at about the same time (1952) that H. and A. Moscona introduced the first procedure to be widely used for the dissociation of cells so that cultures could be established using an initial suspension of individual cells in liquid medium. Their method was to expose embryonic tissues to the action of the enzyme trypsin in a calcium and magnesium free (CMF) saline (these ions promote the adhesion of cells to each other). The dissociated cells were then resuspended in a nutrient medium.

As tissue culture methods became diversified and applied more and more widely, it became necessary to produce exact definitions of the terms which had come into general use in referring to particular aspects of the technique. Hence a Committee on Nomenclature was set up, which produced a list of recommended definitions in 1966. This has recently been revised and brought up to date (Schaeffer, 1979); the following are the recommended definitions of some of the more commonly used expressions, in alphabetical order.

cell culture – this term is used to denote the growing of cells *in vitro*, including the culture of single cells. In cell cultures, the cells are no longer organised into tissues.

cell line – a cell line arises from a primary culture at the time of the first subculture. The term implies that cultures from it consist of numerous lineages of cells present in the primary culture.

cell strain – is derived either from a primary culture or a cell line by the selection or cloning of cells having specific properties or markers. The properties or markers must persist during subsequent cultivation.

chemically-defined medium – a nutritive solution for culturing cells in which each component is of known chemical structure.

clone – a population of cells derived from a single cell by mitosis.

explant – tissue taken from its original site and transferred to an artificial medium for growth.

organ culture – the maintenance or growth of organ primordia or the whole or parts of an organ *in vitro* in a way that may allow differentiation and preservation of the architecture and/or function.

passage – the transfer or transplantation of cells from one culture vessel to another. This term is synonymous with the term 'subculture'.

population density – the number of cells per unit area or volume of a culture vessel.

primary culture – a culture started from cells, tissues or organs taken directly from organisms. A primary culture may be regarded as such until it is subcultured for the first time. It then becomes a 'cell line'.

subculture – see 'passage'.

substrain – can be derived from a strain by isolating a single cell or groups of cells having properties or markers not shared by all cells of the strain.

suspension culture – a type of culture in which the cells multiply while suspended in medium.

tissue culture – the maintenance or growth of tissues, *in vitro*, in a way that may allow differentiation and preservation of the architecture and/or function.

The Essential Requirements for Successful Tissue Culture

In the brief outline of the history of tissue culture at the beginning of this chapter, only a few of the many scientists who contributed to the establishment of culture techniques were mentioned. Their work, and that of many others, eventually made it possible to identify the requirements which must be satisfied if living cells are to be successfully cultured *in vitro*. They may be summarised as follows:

Sterility

Harrison had considerable difficulty in avoiding infection in his cultures, and it was obvious from the outset that rigid asepsis was absolutely essential in the handling of tissues and the preparation of apparatus and solutions. The excision of tissue from an animal immediately deprives the explanted cells of the natural protection afforded them *in vivo* against infection by bacteria and viruses. All the glass and metal items used must be sterilised, either in an autoclave or by heating to 180°C for an hour in an oven. Media and solutions cannot, of course, usually be dealt with in this way, but can be rendered free of bacteria and yeasts, the commonest contaminants, by passage through a filter with a pore size of 0.2 μm. Plastic items (dishes, flasks, syringes, etc.) acquired from commercial suppliers will usually have been sterilised by radiation. The risk of infection may be reduced by handling cultures under a hood through which filtered air is circulating.

Even with good aseptic technique, cultures may still occasionally become infected. Hence antibiotics, such as penicillin and streptomycin, and antifungals, such as mycostatin, may routinely be included in the culture medium at a concentration sufficient to inhibit the growth of contaminating organisms without being toxic to the cells.

Correct Osmotic Pressure

Saline solutions and nutritive media must be isotonic with the body fluids in which the cells existed *in vivo* (an osmotic pressure of 7.6 atm. at 38°C for mammalian cells). Inorganic salts, and especially NaCl, are the chief components which determine the osmotic pressure of animal body fluids, and the same is true of the balanced salt solutions and synthetic nutritive media used in tissue culture.

Control of Hydrogen Ion Concentration

Although cells may withstand some fluctuation in the hydrogen ion concentration in the fluid in which they are immersed, they thrive best at levels of pH between 7.2 and 7.4. It must be remembered that living cells, because of their metabolic activity, will change the pH of their nutritive medium (tending to increase its acidity) and it is important to control the speed of this change by ensuring that a buffer is present, usually in the form of $NaHCO_3$ or HEPES (see p. 17).

Provision of Essential Nutrients

The identification of the substances which are indispensable to the continuing life of cells, as opposed to those which are merely beneficial or even unnecessary, is perhaps the most difficult problem to have been encountered in the evolution of tissue culture technique. The early workers used natural body fluids such as plasma which automatically met all the requirements of the cell, but the precise constitution of these fluids could not be known. Since the early efforts of W.H. and M.R. Lewis, the ultimate goal has been to devise a synthetic nutritive medium, of exactly defined composition, capable of supporting indefinitely the life of cells *in vitro*. Such a medium must contain the following:

Water. The water in which the various components of the medium are dissolved must be of the highest purity.

Inorganic Ions. These are needed for the maintenance of the correct osmotic pressure in the medium, but are in many cases also essential for biochemical and physiological mechanisms within the cells. They include sodium, potassium, calcium, magnesium, iron, chloride, phosphate, carbonate and possibly sulphate.

Carbohydrate. This is most readily provided in the form of glucose, usually at a concentration of 1.0 gram per litre, although a higher concentration is preferable for the cultivation of nervous tissue.

Amino-acids. There is a difference between the needs of intact animals and those of cells *in vitro* in the number of amino-acids which must be supplied. For animals, certain amino-acids are termed 'essential' because they must be present in the diet, since they cannot be synthesised by the animal. These include arginine, histidine, isoleucine, leucine, lysine, methionine, phenylalanine, threonine, tryptophane and valine. In addition to these, cells *in vitro* should also be supplied with some others, such as glutamine and tyrosine (in the intact animal, these are mainly synthesised in the liver).

Vitamins. The range of vitamins which must be supplied varies to some extent, depending upon the particular type of cell, but there is a general requirement for several vitamins of the B group. These include thiamine, choline, riboflavin, pyridoxal, pantothenic acid, nicotinic acid, para-aminobenzoic acid, biotin, folic acid and inositol, needed chiefly for their involvement as co-enzymes in biochemical reactions.

Additional Factors. There is some uncertainty about the necessity for the provision of a protein or polypeptide component in the medium, and the same is true in the case of purines and pyrimidines, hormones, extra amino-acids (e.g. glycine, serine and proline), and vitamins A and C. Generally speaking, it is probable that these substances, while not absolutely essential for the basic survival and growth of cells *in vitro*, may be useful in promoting the differentiation of particular types of cell, especially in organ cultures.

Oxygen and Carbon Dioxide. Although cells may be able to satisfy their energy requirements in the absence of oxygen by anaerobic glycolysis, they are unlikely to survive indefinitely without sufficient oxygen to maintain the Krebs cycle. Conversely, an excessively high oxygen tension in the medium may be toxic to the cells. Carbon dioxide is difficult to eliminate from a culture, since it is produced by the cells; in any case, it is probably essential as a source of carbon and for some biochemical processes. It also has an important role in the control of the pH of the culture medium when $NaHCO_3$ is used as the main buffer.

Temperature

As a general guide, cells will grow and maintain their specific functions best at the body temperature of the animal from which they were taken (37-38.5°C for mammalian and avian cells). Temperatures above the optimum level will almost certainly damage or kill the cells; subnormal temperatures are less deleterious, but they will inhibit growth and slow down cellular activities.

Tissue Culture Media

Since this is not a monograph on tissue culture technique, our aim is to provide no more than a general outline of culture media and methods, with emphasis on those which are particularly relevant to research on cell locomotion. Detailed guidance is obtainable in the books recommended for further reading at the end of this chapter.

It is usual to divide tissue culture media into two categories, natural and defined (or synthetic). The former consist of body fluids or tissue extracts from animals; they are relatively easy to produce and can usually be relied on to provide all the factors required for the successful cultivation of cells, but they represent an unknown and uncontrolled variable in research. Synthetic media are much more difficult to produce, but offer the possibility of studying living cells in an environment which, initially at least, is fully defined and can be reproduced repeatedly; this statement has to be qualified because the metabolism of the cells will cause changes in the composition of the medium which cannot be predicted exactly. It also has to be borne in mind that there are only a few cell strains which can be maintained indefinitely in a totally defined medium, and in most cases it is necessary to add 5-10 per cent of a natural medium (usually serum) to a synthetic medium to guarantee healthy and persisting growth.

Natural Media

Clotted plasma was one of the first media to be introduced in the early days of tissue culture, and proved so effective that, for a considerable time, it was the most commonly used medium. Although it has largely been superseded, it continues in use to a limited extent. Its advantages are that, as well as providing nutrients, it will hold explants in position on a glass or plastic surface, and it contains a three-dimensional meshwork of fibrin strands which can support the movement of cells in any direction. Cockerel plasma is particularly suitable, since it forms a firm transparent clot; small quantities are obtained by withdrawing blood aseptically from a wing vein, using a syringe and collecting tube coated with a solution of the anticoagulant, heparin. After centrifugation, the clear liquid plasma is transferred to a refrigerator for storage. Larger quantities of plasma can be collected by cardiac puncture or by inserting a cannula into the carotid artery in the neck.

Like plasma, embryo extract is no longer in general use as a tissue culture medium, but continues to have restricted application as an extremely rich source of nutrients. The simplest way of preparing it is to incubate fertile hen eggs for not more than ten days, remove the embryos aseptically, and put them in the barrel of a sterile 20 ml syringe; the plunger is then inserted and the embryos are pushed through the nozzle. This produces a thick 'soup' of embryo tissue, which is collected in a sterile tube and diluted with an equal volume of balanced salt solution. The mixture is then centrifuged, and the clear supernatant, free of tissue fragments, is decanted and may be stored in a deep freeze.

Serum is now the most extensively used natural medium, generally in the proportion 10 per cent serum:90 per cent synthetic medium. There is no necessity for the serum to be taken from the same animal as the cells, or even from the same species. Human, horse or calf serum is most often used, and serum from young donors is preferred because it promotes better growth. Thus human placental serum, prepared from blood collected from the placenta via the umbilical cord after delivery of the infant, or fetal calf serum, may be recommended for the cultivation of some tissues (e.g. nervous tissue). For most purposes, however, new-born calf serum is perfectly suitable.

Synthetic Media

Synthetic media may be said to have three levels of complexity; the simplest, known as 'balanced salt solutions', contain only inorganic ions and glucose, in which cells will survive for no more than a few hours. They are useful for washing cells or fragments of tissue, and for their short-term maintenance during procedures such as dissociation with trypsin (when a saline solution devoid of calcium and magnesium is often used), but they are also basic to all synthetic media.

At the second level of complexity, there are media which, without any additions, supply a sufficient range of nutrients to allow the survival of cells for several days, but for indefinite growth these must be supplemented with 5 or 10 per cent of serum. The majority of the research on cell locomotion *in vitro* has been carried out on cells cultured in media of this type.

The media with the greatest number of components are those which have been designed to support the long-term growth of cells in the absence of any natural and undefined additive. There are now a few cell lines which have been constrained to grow indefinitely in these totally defined media, but they are not particularly applicable to the investigation of cell locomotion and will not be discussed here.

Balanced Salt Solutions. There are at least half-a-dozen different recipes for balanced salt solutions, and there is little to choose between them. One factor which has to be considered, however, is the nature of the gas mixture which the cells will be exposed to *in vitro*; this may be air or 5 per cent CO_2 in air, and in the latter case a higher concentration of the buffer, $NaHCO_3$, is needed. The composition of two of the most frequently used balanced salt solutions is shown in Table 2.1; phenol red is included to give a visual indication of changes in pH.

In recent years an alternative buffer has been introduced. This is N-2-hydroxyethylpiperazine-N′-2-ethanesulphonic acid or 'HEPES', which has the advantage that it operates at its highest efficiency in the appropriate pH range, i.e. at pH 7.3.

Eagle's Minimum Essential Medium. This is an example of a synthetic medium of the second level of complexity, one of several similar formulations which are frequently chosen for work on cell locomotion. Its composition is set out in Table 2.2, but the ingredients of the basic balanced salt solution, which may be Earle's or Hanks', are omitted. It can be seen to consist of only 21 items; the 13 amino-acids include ten 'essential' ones plus cystine, glutamine and tyrosine, the last two for the reason mentioned earlier and the first one because cells *in vitro* tend to lose cystine into the medium when they have synthesised it. There is some doubt about the stability of glutamine, which should preferably be added to the medium just before it is used. The eight vitamins are all of the B group.

Like most standard media, Eagle's medium is commercially available, either ready for use or in concentrated or powdered form. The necessary serum (e.g. new-born calf) can also be bought, presterilised and checked by the supplier for lack of toxicity. Liquid synthetic media may be stored at 4°C in a refrigerator, but sera and solutions of glutamine are best preserved at –20°C in a deep freezer.

Table 2.1: Composition of Earle's and Hanks' balanced salt solutions (grams per litre) (from Paul, 1975)

	Earle's saline	Hanks' saline
NaCl	6.80	8.00
KCl	0.40	0.40
$CaCl_2$	0.20	0.14
$MgSO_4,7H_2O$	0.10	0.10
$MgCl_2,6H_2O$	—	0.10
NaH_2PO_4,H_2O	0.125	—
$Na_2HPO_4,2H_2O$	—	0.06
KH_2PO_4	—	0.06
Glucose	1.00	1.00
Phenol Red	0.05	0.02
$NaHCO_3$	2.20	0.35
Gas phase	5% CO_2 in air	Air

Table 2.2: Composition of Eagle's minimum essential medium (mg per litre) (from Paul, 1975)

Amino acids		Vitamins	
L-Arginine HCl	126.4	D-Ca-Pantothenate	1.0
L-Cystine	24.0	Choline chloride	1.0
L-Glutamine	292.0	Folic acid	1.0
L-Histidine HCl,H_2O	41.9	i-Inositol	2.0
L-Isoleucine	52.5	Nicotinamide	1.0
L-Leucine	52.4	Pyridoxal HCl	1.0
L-Lysine HCl	73.1	Riboflavine	0.1
L-Methionine	14.9	Thiamine HCl	1.0
L-Phenylalanine	33.0		
L-Threonine	47.6		
L-Tryptophan	10.2		
L-Tyrosine	36.2		
L-Valine	46.8		

Surfaces for the Establishment of Monolayers of Cells

Dissociated cells can be cultured in bulk by suspending them in a relatively large volume of liquid medium in a flask or bottle, but they are then not readily accessible for examination under the microscope. The optimum conditions for microscopy can only be provided when the cells are grown as a monolayer on the surface of a suitably thin layer of transparent material, which must be non-toxic and scrupulously clean. The materials most often used are, of course, glass or plastic, the first in the form of a coverslip and the second constituting the floor of a plastic Petri dish or small flask with parallel sides. When coverslips are used, they must undergo thorough cleaning, with either a suitable detergent or strong acid, followed by repeated rinsing in distilled water, prior to sterilisation with dry heat. Plastic culture dishes or flasks specifically for tissue culture from a commercial supplier will already be perfectly clean, and are normally used only once and then discarded.

Many types of cell will adhere to and spread on a glass or plastic surface, but in some cases there may be advantages in covering the surface with a layer of collagen gel. One way of preparing the gel is to dissolve the tendons from rats' tails in acetic acid. The solution is dialysed against distilled water to remove the acid, and a drop of the viscous solution is spread over the surface on which cells are to be grown, and allowed to dry. Thus a thin layer of collagen fibres is provided, which the cells find more congenial than naked glass or plastic and to which they will more readily adhere. By incorporating a suspension of separated cells into a thicker layer of collagen gel, it is also possible to study the movements of cells in a three-dimensional matrix which more closely resembles their normal environment *in vivo* (Schor, 1980).

An alternative method of providing a more suitable surface for monolayer cultures is to coat the glass or plastic with a polymerised amino-acid (e.g. polylysine).

Culture Techniques for the Study of Cell Locomotion

The Hanging Drop Technique

This is the oldest technique in tissue culture, used by Harrison in his experiments on the growth of nerve fibres in 1907, but it is still preferred for the cultivation of some varieties of cells and tissues, and has been employed in many studies on cell locomotion.

The basic equipment consists of a glass coverslip and a thick glass slide with a concavity on one surface. An explant of tissue is placed in a drop of liquid cockerel plasma in the centre of the coverslip and mixed with a roughly equal volume of chick embryo extract. When the plasma has clotted, the coverslip is inverted over the concavity in the slide, and sealed in position with molten paraffin wax painted along its edges to prevent the evaporation of water from the culture (Figure 2.2). Alternatively, a drop of liquid medium (synthetic plus serum) may be used instead of a plasma clot, but particular care is then needed to keep the fluid in the centre of the coverslip when handling the cultures.

Figure 2.2: Diagram showing a sectional view of the hanging drop culture technique

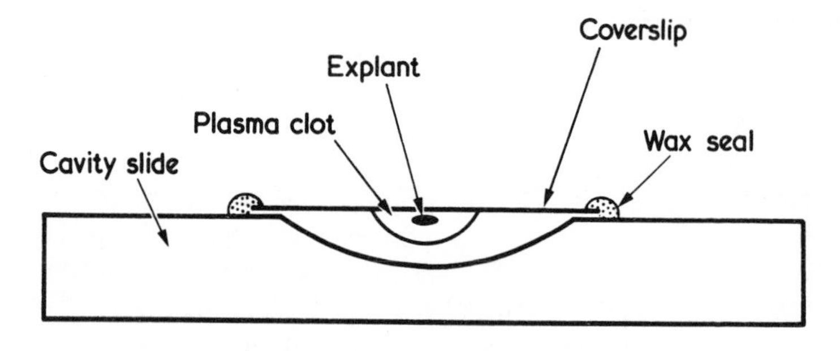

The advantages of hanging drop cultures are that they require only simple and relatively inexpensive materials and that the cells emerging from the explant on to the coverslip can be examined microscopically. However, the volume of nutrient medium available to the cells is small, and it is difficult to avoid infecting the cultures when the coverslip is removed from the slide to replenish the medium. It is easier to keep the culture sterile in the Maximow 'double coverslip' version of the hanging drop; here the coverslip carrying the culture is smaller than the diameter of the concavity and is attached to a larger coverslip by the surface tension exerted by a drop of sterile saline spread between the two. The larger coverslip is then sealed to the slide with wax as before (Figure 2.3). Thus when the culture requires to be fed, the smaller coverslip can be detached from the larger one with sterile forceps, the medium replenished, and the culture transferred to a clean, sterile, large coverslip and slide, with less risk of infection.

One disadvantage of the hanging drop technique results from the curvature of the concavity in the thick glass slide, which impairs the image produced by the phase contrast microscope. Consequently, if time-lapse cinemicrography is to be used to record the movements of the cells, the coverslip may have to be detached from the slide and incorporated into a chamber with parallel surfaces which will allow the phase contrast microscope to be used. Such a chamber can easily be constructed using a thin plate of non-toxic stainless steel or plastic with a central hole; the coverslip bearing the culture is sealed on to one surface of the plate, with the hanging drop occupying the middle of the hole, and a clean coverslip sealed on to the other surface completes the chamber and converts the drop into a column of liquid in contact with both coverslips.

Figure 2.3: Diagram showing a sectional view of the Maximow double coverslip culture technique

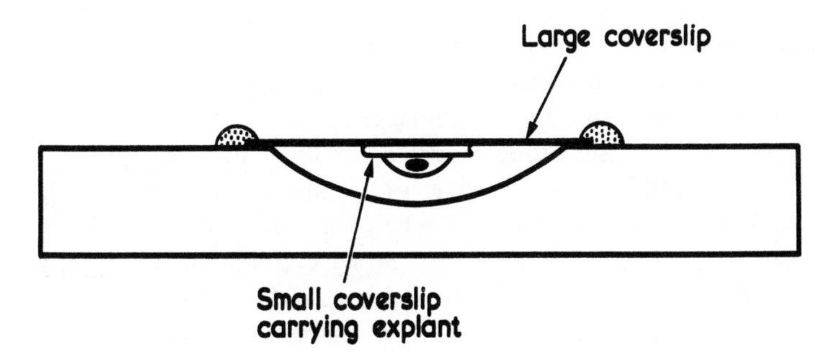

Dissociation Methods

When explants (excised fragments of tissue) are used to establish cultures, the locomotory behaviour of the cells which emerge from the explants can be observed, but this procedure has disadvantages if the object of the experiment is to study the movements of initially separate cells. It is much better to use cultures established by producing a suspension of individual dissociated cells. The cells are then available for study as soon as they have settled on to the floor of the container, and, provided that an appropriate number of cells per unit area of substratum is introduced into the container, they will be separated from each other by an average distance which is adequate to allow them to move initially in any direction but small enough for collisions to occur so that the effects of cell-cell contact can be observed. Incidental advantages of using dissociated cells include the possibility of working with established strains of cells, and the relative ease with which a uniform population of cells can be maintained and subcultured.

Satisfactory methods for the dissociation of cells were introduced in the early 1950s, based on the action of the enzyme trypsin, and this procedure is still widely used. The technique, in outline, is as follows. Pieces of tissue are cut into small fragments and rinsed in a balanced salt solution devoid of calcium and magnesium ions. They are then placed in a solution of trypsin in the same salt solution, in a small flask which may be agitated for an appropriate time (depending on the nature of the tissue). The cells are gently separated by repeatedly drawing the solution into a pipette and expelling it back into the flask. The suspension of detached cells is then collected and centrifuged, the supernatant trypsin solution discarded, and the cells resuspended in culture medium with serum. The number of cells per unit volume in the suspension is counted, using a haemocytometer (or an automated cell counter), and the concentration adjusted to produce a final population density appropriate for the particular type of cell, and for the size of the container in which they are to be grown.

The most popular type of container for dissociated cell cultures is the plastic Petri dish; dishes prepared for tissue culture, and presterilised, are available from a number of commercial sources. They have loose-fitting lids and must therefore be put in a humidified incubator to prevent loss of water from the medium by evaporation. The incubator is usually equipped with a device which monitors the CO_2 in the contained air and injects sufficient CO_2 to maintain the concentration at 5 per cent if necessary.

Although trypsin is the enzyme most often used for the dissociation of cells, other enzymes are sometimes employed, either alone or in combination with trypsin. Thus collagenase may be useful in obtaining cells from tissues with a high collagen content; papain (a mixture of enzymes, from a plant, which digest extra-cellular proteins) or pronase (proteolytic enzymes derived from cultures of *Streptomyces griseus*) may also be used.

In some instances it may be possible to disaggregate cells by removing divalent cations (particularly calcium and magnesium) by means of a chelating agent such as ethylenediaminetetra-acetic acid (EDTA or Versene). The choice of enzyme or chelating agent, alone or in combination, depends on the nature of the organ or tissue to be dissociated. Much of the work on cell locomotion has involved the study of the 'fibroblast' cell type, monolayer cultures of which can be prepared by dissociating a variety of different tissues. Epithelial cells have been used less often; quite pure populations of such cells may be obtained by using an enzyme or chelating agent initially to detach a layer of surface epithelium *en masse* from its underlying connective tissue, with subsequent dissociation of the individual cells from the sheet.

When monolayer cultures have been established, an enzyme or chelating agent can also be used to detach the cells from the Petri dishes for subculturing; they are then resuspended in culture medium and transferred to new dishes for continued maintenance and growth.

Figure 2.4: Diagram showing a sectional view of the Cooper culture dish

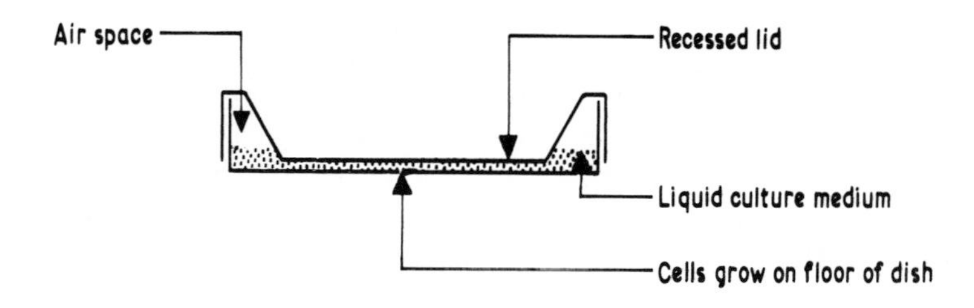

While ordinary plastic Petri dishes are adequate for routine microscopic examination and photography of the cells at low magnification, they are not ideal for photomicrography or cinematography at higher magnifications, firstly because the distance between the lid of the dish and its floor may be too great to allow the condenser of the microscope to be properly focused (unless a special long working distance condenser is used), and secondly because the plastic floor of the dish may impair the quality of the image produced by higher power objectives. The first difficulty can be overcome by growing the cells in a specially designed dish (Cooper, 1961) which has a recessed lid (Figure 2.4). The second may require a different approach, in which the cells are allowed to settle on and adhere to glass coverslips placed on the floor of a dish; a coverslip can then be removed and incorporated into a chamber of the type described on p. 21, which provides the optimal conditions for phase contrast microscopy and will allow an oil immersion objective to be used.

Fibroblasts and Epithelial Cells in Culture

In 1960, E.N. Willmer classified the cells most often seen in tissue cultures as follows:

The emergent cells can, in general, be grouped into three or four main categories according to their appearance, type of movement, staining reactions and general growth-habits and requirements. These main categories are: epitheliocytes, mechanocytes (fibroblasts) and amoebocytes (wandering cells).

Although he was referring to cells emerging from explants of tissue, Willmer's classification remains applicable to cultures of dissociated cells established as monolayers on a suitable substratum. In the context of the study of cell locomotion, little use has so far been made of Willmer's third category, amoebocytes, although this would include the most rapidly motile cells in the body such as lymphocytes, monocytes or macrophages, and polymorphonuclear leucocytes (PMNs). This is unfortunate, since the members of this group are of prime importance in the establishment and maintenance of immunity; however, some work on PMNs is referred to in Chapter 7 (page 147).

Figure 2.5: Chick embryo heart fibroblasts cultured for 24 hours, fixed, and stained with Harris's haematoxylin

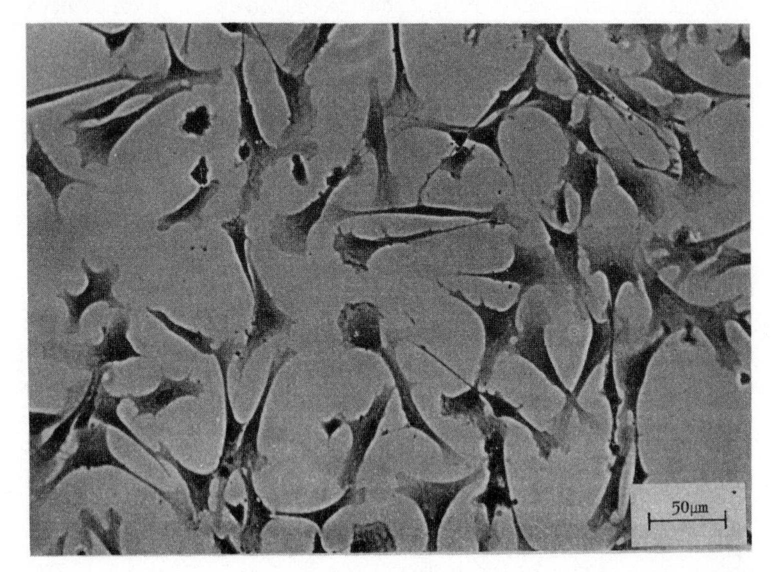

It is the locomotory behaviour of the fibroblast which has been studied most intensively. This cell has a number of attributes which make it particularly useful for work on cell movement. It is relatively

easy to establish and maintain *in vitro* and it can be obtained from most of the tissues of the body. When fibroblasts have settled on a suitable substratum they tend to adopt an elongated, angular shape, with a number of short processes projecting from the main mass of cytoplasm (Figure 2.5). As they make contact with each other via these processes, they proceed to form an open network. A fibroblast which is actively motile often displays a characteristic polarised outline; the cell will often be triangular, with a tapering process extending from the 'posterior' end and leading lamellae (ruffled membranes) at the 'anterior' end (Figure 2.5).

The locomotion of epithelial cells has recently attracted more attention because of the importance of such cells in wound healing and their frequent involvement in cancer. *In vivo*, epithelial cells commonly adhere to each other very closely in order to carry out their basic functions of covering body surfaces and forming glands; in doing so they frequently produce specialised areas of contact such as the desmosome (macula adhaerens) and the tight junction (zonula occludens).

Figure 2.6: Epithelial outgrowth from an explant of chick embryo gizzard, fixed, and stained with Harris's haematoxylin

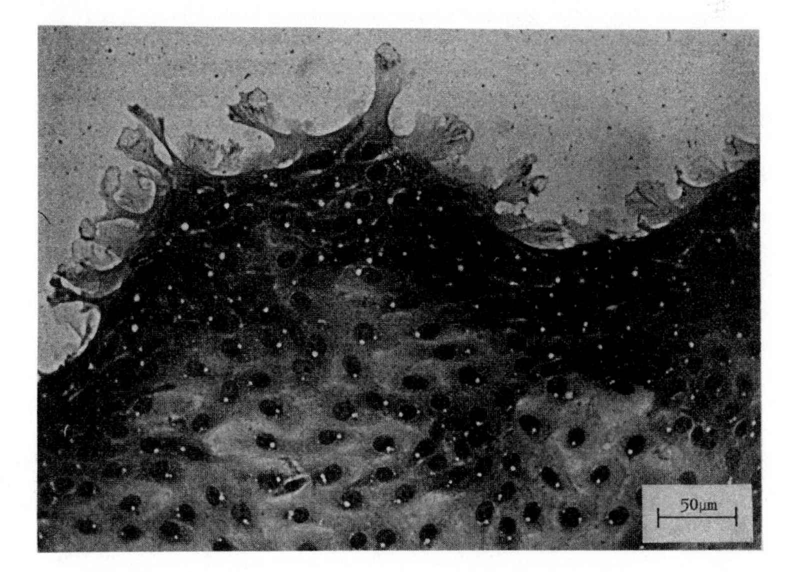

This pattern of behaviour is carried over into the tissue culture environment, where monolayers of epithelial cells adhere to form

extensive continuous sheets (Figure 2.6). Within such sheets the individual cells generally have a regular outline, abutting closely against the edges of their neighbours, and forming tight junctions and desmosomes. Epithelial cells which find themselves at the edge of a sheet, however, usually display a more variable shape, and the unattached portion of their periphery forms leading lamellae (Figure 2.6).

References

A. Carrel and M.T. Burrows (1911) 'Cultivation of Tissues *In Vitro* and Its Technique', *J. exp. Med.*, vol. 13, p. 387

W.G. Cooper (1961) 'A Modified Plastic Petri Dish for Cell and Tissue Cultures', *Proc. Soc. exp. Biol. Med.*, vol. 106, p. 801

R.G. Harrison (1910) 'The Outgrowth of the Nerve Fiber as a Mode of Protoplasmic Movement', *J. exp. Zool.*, vol. 9, p. 791

M.R. Lewis and W.H. Lewis (1911) 'The Cultivation of Tissues from Chick Embryos in Solutions of NaCl, $CaCl_2$, KCl and $NaHCO_3$', *Anat. Rec.*, vol. 5, p. 277

W.I. Schaeffer (1979) 'Proposed Usage of Animal Tissue Culture Terms (Revised 1978)', *In Vitro*, vol. 15, p. 649

S.L. Schor (1980) 'Cell Proliferation and Migration on Collagen Substrata *In Vitro*', *J. Cell Sci.*, vol. 41, p. 159

E.N. Willmer (1960) *Cytology and Evolution* (Academic Press, London), p. 18

Suggestions for Further Reading

J. Paul (1975) *Cell and Tissue Culture*, 5th edn. (Churchill Livingstone, Edinburgh)

P.F. Kruse and M.K. Patterson (eds.) (1973) *Tissue Culture, Methods and Applications* (Academic Press, London and New York)

J.A. Sharp (1977) *An Introduction to Animal Tissue Culture* (Institute of Biology, Studies in Biology No. 82, Edward Arnold, London)

3 LIGHT AND ELECTRON MICROSCOPY OF CULTURED CELLS

The human eye can detect variations in the amplitude (brightness) and in the wavelength (colour) of the light which forms an image on the retina. When a histological section is examined through a light microscope, the cells and intercellular materials in the tissue are made visible by virtue of the contrast produced by the dyes used to stain the section; these absorb some wavelengths but transmit others, giving a coloured image.

When a light microscope with conventional optics is focused on a monolayer culture, the living cells are scarcely visible, since there is little difference in amplitude between light passing through a cell and light passing through the surrounding medium, and hence little inherent contrast (Figure 3.1a). Histological stains cannot be used to introduce contrast, as this would inevitably kill the cells.

During the early years of tissue culture, some use was made of dark ground illumination, which renders the edge of a cell and of its nucleus, and cytoplasmic structures such as vacuoles and mitochondria, visible as relatively bright lines against a dark background (Figure 3.1b). But in this system only a minute proportion of the light directed obliquely through the specimen by the condenser passes into the objective and even with a powerful source of light the image is faint and long exposures are needed for photomicrography. An alternative which was used was to close down the substage iris diaphragm and slightly defocus the microscope, which enhanced the contrast in the image, but at the cost of some loss of resolving power and the creation of potentially misleading diffraction effects.

The Phase Contrast Microscope

It was not until the 1940s that a more satisfactory optical system for the detailed study of living cells *in vitro* became available in the form of phase contrast microscopy. This relies upon the fact that light which is diffracted by structures in the specimen which are optically denser than their surroundings is retarded, that is, its wave pattern is no longer in phase with that of light passing straight through the specimen without

Figure 3.1: Photomicrographs of a living culture of fibroblasts using (a) bright field, (b) a dark ground condenser, (c) phase contrast, (d) Nomarski interference contrast

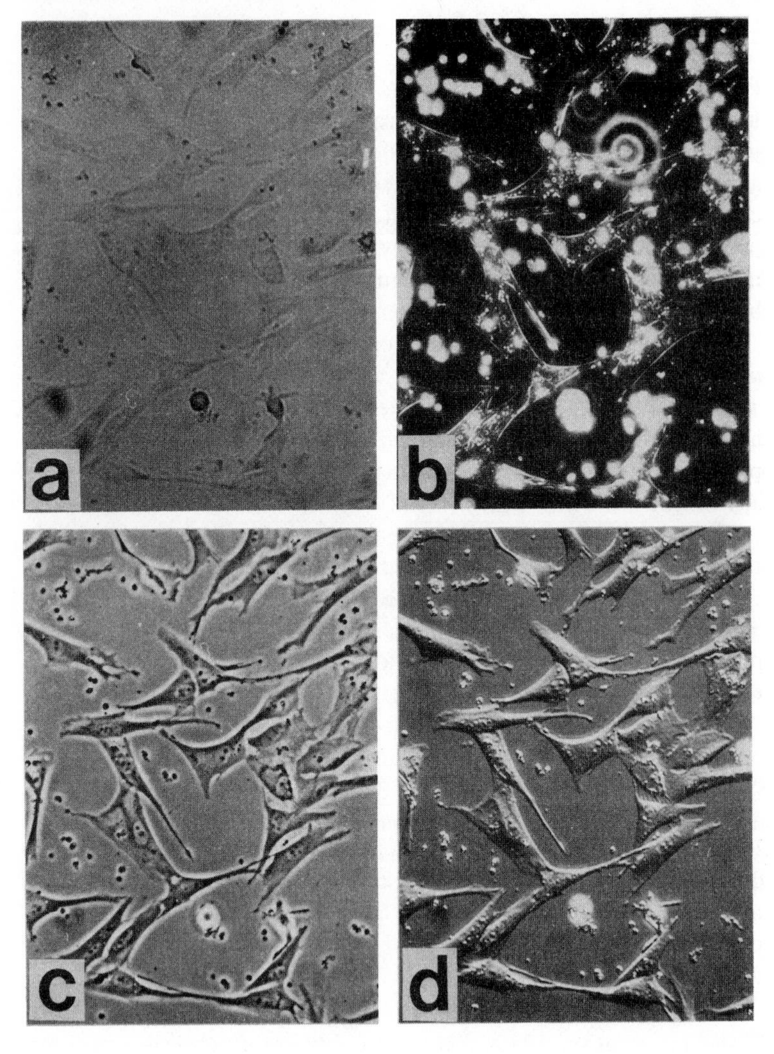

being deviated. A device called a phase plate, placed behind the rear lens of the microscope objective, has an annular groove, whose floor is coated to absorb light. The diffracted rays, which pass through the thicker regions of the phase plate, are further retarded in comparison with the undiffracted rays, which are directed through the thinner, coated floor of the groove (and thus reduced in amplitude). Subsequent interference between the out-of-phase waves of the diffracted and the undiffracted light produces an image in which optically dense structures appear (usually) darker than their surroundings (Richards, 1954). The cells (e.g. in a monolayer) are thus made visible, the chromatin and nucleoli within the nucleus appearing relatively dark, as do granules and mitochondria in the cytoplasm; cytoplasmic vacuoles, on the other hand, will appear relatively bright (Figure 3.1c).

In addition to the special phase contrast objectives, the microscope must have a condenser incorporating a series of phase diaphragms, each with a transparent annulus of different radius; these are usually built into a rotating disc so that the diaphragm with an annulus of appropriate size for each objective can be positioned below the condenser aperture.

These additional components do not impair the resolving power of the microscope significantly, and, although there is considerable absorption of light, the image is still bright enough to allow reasonably short exposures for photomicrography. The chief undesirable feature of the phase contrast image is the presence of a bright halo round the edge of each cell, which may obscure details in this region.

Nomarski Interference Contrast

Introduced in the 1950s, this generates an image of living cells quite different from that produced by phase contrast. The optical system is more complex, and the microscope requires several additional components, including a polariser and a series of double quartz prisms (a different one for each objective) in the condenser, plus an adjustable double quartz prism and an analyser in the body behind the objective. Because polarised light is used, strain-free objectives without birefringence give the best results.

By adjusting the double quartz prism in the body of the microscope, a coloured or a monochrome image can be selected, the latter usually being preferable (Figure 3.1d). One's first impression when looking at a Nomarski interference contrast image is that it is similar to the surface view seen in a scanning electron microscope; this is deceptive, since the image is not formed by light reflected from the cell surface but by light which has passed through the cell. Hence all the structures within the cell which differ sufficiently in optical density from their surroundings are made visible, but some (e.g. nucleoli and cytoplasmic granules) appear to be convex and others (e.g. vacuoles) concave (Padawer, 1968).

When compared with phase contrast, the Nomarski image lacks the potentially troublesome halo around the cell border, but it has no major advantages; the two systems should be regarded as complementary, and it may be useful to have a microscope equipped with both, so that the same feature can be examined or photographed with phase and with Nomarski optics.

Interference Reflection Microscopy

Phase contrast and Nomarski interference contrast allow the general appearance and behaviour of living cells to be studied; interference reflection microscopy is designed to provide information about a particular feature of a cell *in vitro*, the relationship between its membrane and the surface on which the cell is moving. It was first used for this purpose by Curtis (1964), who adapted an existing optical device for measuring the thickness of thin films, and the technique was re-evaluated and refined by Izzard and Lochner (1976).

A microscope equipped for incident light illumination of the specimen is required. In such a microscope the illumination passes through the objective, which acts as a condenser focusing it on the specimen (see Figure 3.5). Light which is reflected from the specimen back through the objective is then used to form an image. Interference reflection microscopy was developed by Curtis (1964) on the assumption that the layer of medium between the cell and the coverslip over which it was moving would behave as a thin film, generating an interference pattern in the light reflected from it. Differences in the thickness of this film in different areas can be detected as differences in the interference colours (with white light) or as differences in intensity (with monochromatic light). The thickness of this film of medium depends on the separation between the lower surface of the cell and the coverslip, and the image therefore provides information as to the closeness of the contact between the cell and its substratum (see Figure 6.7b). In practice the images obtained with the interference reflection microscope are often more complex than might be expected from these considerations, due apparently to the fact that the cell itself can act as an additional thin film which also contributes to the image (Izzard and Lochner, 1976). This difficulty can be overcome to some extent by an optical system that allows a high illuminating numerical aperture to be used (Izzard and Lochner, 1976). Even so, in regions where the cell is very thin (1.0 μm or less) the second layer constituted by the cell can still make a significant contribution to the image (Gingell, 1981). In general, however, regions where the cell approaches very closely to the coverslip appear darker than regions where the two are more widely separated, and interference reflection microscopy has provided useful information concerning the nature, distribution and behaviour of contacts between a cell and the surface on which it moves.

Time-lapse Cinemicrography

The speed of movement of individual cells *in vitro* varies a good deal. There are differences in the rates of locomotion achieved by different types of cell. Thus some epithelial cells have been estimated to have a mean velocity of only 0.12 μm per minute (Middleton, 1973) whereas the mean for lymphocytes is 3.7 μm per minute (Nunn, Sharp and Kimball, 1970). These are average figures, and the actual velocity of a particular cell may change from moment to moment, varying from zero up to 4.0 μm per minute or more. At such speeds living cells, when watched through the microscope, scarcely appear to be moving at all, and many hours of tedious observation would be needed to follow and record (by drawings or 35 mm photomicrographs) the locomotory behaviour of the cells in the field of view. Fortunately, the cine camera can relieve the human observer of the burden of recording cellular activities *in vitro*, and do it more reliably. If the camera, instead of running continuously, is controlled so that there is an interval of several seconds between each exposure, the velocity of the cells is increased many times when the film, after development, is projected at the standard rate of 24 frames per second, and the events which occurred in the field of view of the microscope over a period of several hours are condensed into a film lasting only a few minutes.

Although time-lapse cinemicrography was applied to cells and tissues *in vitro* quite soon after culture techniques first began to be used, it was only in the 1950s, when the phase contrast microscope became available, that the procedure achieved its full potential and the advantages to be gained from its use were soon widely recognised. In addition to those already mentioned, these advantages include the provision of a permanent record of the behaviour of a group of cells, which can be viewed repeatedly; if the film is run through an analysing cine projector, it can be seen in forward or reverse motion, at slow speed, or stopped to allow any individual frame to be examined in detail. Of particular importance is the scope for applying quantitative methods to time-lapse films; by projecting individual frames and outlining the shape and position of the same cell (or cells) at known intervals of time, changes in the area of the cell and in its velocity and direction of locomotion can be measured accurately and, if necessary, subjected to computer analysis.

The three essential pieces of equipment required for time-lapse cinemicrography are the microscope, the cine camera and the camera control unit.

The Microscope

A microscope of the 'inverted' type, with the lamp and condenser above the stage and the objective below, is essential for the examination or filming of cells grown in a Petri dish or flask; the cells are attached to the floor of the container, covered by a layer of liquid medium, and the dish or flask cannot be turned over. Culture chambers consisting of two parallel coverslips separated by a shallow cavity full of medium can be reversed (so that the coverslip carrying the cells is uppermost) and examined with a conventional microscope, but some cells (e.g. lymphocytes) do not adhere very firmly to the coverslip and may detach and sink out of focus. Hence even when using culture chambers of this type, it may still be preferable to use an inverted microscope (Figure 3.2).

The microscope should be equipped with phase contrast (or interference contrast) optics and, if possible, the condenser should have a long working distance so that it can be focused through the full depth of a Petri dish. The body of the microscope should be fitted with a single photo-tube in addition to binocular viewing eyepieces, so that all the available light can be directed either to the camera for filming, or to the eyepieces for routine examination of the culture.

Some means must be provided for keeping the culture at 37°C on the stage of the microscope. This may be done by enclosing the microscope in a perspex box, with the controls extended through openings to the outside; the interior of the box can be maintained at 37°C by warm air blown in from an ordinary hair dryer, controlled by a temperature sensor and thermostat. Alternatively, only the stage of the microscope (and, indirectly, the culture chamber) may be warmed, but there is a tendency for this to create convection currents in the culture medium. Yet another approach is to direct a stream of warm air over the culture chamber. A thermistor may be attached to the surface of the chamber and used, via an electronic proportional control circuit, to provide continuous adjustment of the temperature of the air flowing over the chamber.

Figure 3.2: Equipment for time-lapse cinemicrography. (a) thermo-circulator keeping microscope stage at 37°C; (b) 16 mm cine camera with electric motor; (c) beam splitter with viewing eyepiece and photocell; (d) inverted phase contrast microscope; (e) exposure meter; (f) camera control units

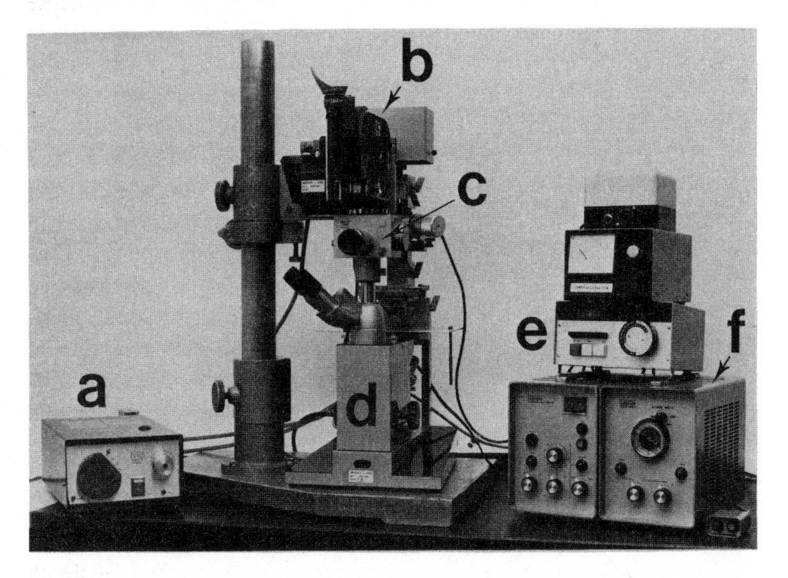

The Cine Camera

A 16 mm camera is generally used because of the good quality of the image produced by 16 mm film. Any commercially available camera is suitable, provided that it is reliable, not prone to damage the film, and is capable of being controlled so that both the length of the exposure and the duration of the interval between exposures can be varied. A certain amount of vibration is likely to be engendered by the camera's mechanism, and it is most important to minimise this and prevent it from being transmitted to the microscope. This can usually be achieved by bolting the camera firmly to a substantial metal pillar with a massive base plate (Figure 3.2), and ensuring that there is no direct contact between the camera and microscope.

A beam-splitter, interposed between the microscope ocular and the camera lens, allows the majority of the light to pass into the camera and directs a small proportion to a viewing eyepiece which allows the field of view and focus of the microscope to be checked during filming. The beam-splitter usually incorporates a photo-cell to enable the intensity of the illumination of the specimen to be monitored.

The Camera Control Unit

This contains electro-mechanical or electronic devices which determine the length of the exposure and the duration of the interval between exposures. The former is dependent on the sensitivity of the film and the amount of light transmitted through the specimen. The duration of the interval will be decided after taking account of factors such as the velocity of movement of the cells being filmed, the length of time for which the culture is to be filmed and the capacity of the film magazine. The control unit must deliver pulses to the camera with accuracy and consistency of timing; if, for instance, there are variations in the length of the exposure, the density of the negative will vary correspondingly from frame to frame on the film.

Apart from this basic function, the control unit will usually be fitted with an outlet which will either switch the microscope lamp on just before the film is exposed and extinguish it when the exposure is completed, or will similarly open and close a light shutter placed between the microscope lamp and the condenser. The second alternative is best when the lamp is of the quartz-halogen type, since these have a rather short life when they are switched on and off frequently. This precaution is necessary to prevent damage to the cells in the culture by over-exposure to the quite intense radiation produced by the microscope lamp.

Time-lapse Video-tape Recording

This procedure involves the use of a television camera and video-tape recorder, the latter having either a single shot facility capable of recording single scans at a pre-set interval (determined by an intervalometer), or being run continuously at a slow speed, with playback at normal speed. In either case the speed of movement of the cells is increased, just as it is with time-lapse cine film, when the recording is viewed on a television monitor screen.

The advantages of this method of recording cell behaviour include lower running costs (the tape is relatively cheap, requires no photographic development or printing, and can be used more than once), and immediate viewing (no delay for development). By employing a very sensitive television camera, recording is possible with minimal levels of illumination of the culture.

The disadvantages stem from the comparatively poor resolution available on the television screen as compared with cine film, together with the curvature of the screen which makes accurate measurement of distances within the image difficult.

Immunocytochemistry and Fluorescence Microscopy

The contribution of immunocytochemistry to the study of cell locomotion has been invaluable because it has made it possible to reveal the intracellular distribution of proteins which are of fundamental importance in cytoplasmic movement. The genesis of the technique may be traced to 1934, when Marrack conjugated anti-typhoid and anti-cholera sera with a red dye, and demonstrated that the chemically modified antibodies were still capable of combining specifically with the homologous organisms, which were thereby coloured red. Similar experiments were later conducted by Coons, using pneumococcal antibody. He confirmed that the chemical combination of an antibody with a dye did not impair its ability to react specifically with the appropriate antigen, but found that the colour imparted by the dye was not sufficiently intense to be of much practical use. It occurred to Coons that a fluorescent compound might be easier to detect, and he therefore tried the effect of conjugating anthracene with an antibody (Coons *et al.*, 1941). The result was encouraging, since the organisms to which the labelled antibody became attached fluoresced brightly when exposed to ultraviolet light, but unfortunately the anthracene gave a blue fluorescence similar to that produced by the natural autofluorescence of the tissues in which the organisms were lying. An important step was taken in the following year when Coons and his collaborators overcame this problem by combining antibody with the dye fluorescein; this gave bright green fluorescence in ultraviolet light, contrasting well with the blue autofluorescence of the tissues (Coons *et al.*, 1942). Thus the basis was established for a technique which has become a standard method for the tracing of specific proteins in cells or tissues.

Although the procedure has been improved in detail, it remains essentially the same today, and depends upon two phenomena: firstly, the specificity of the reaction between an antibody, produced in an immune response, and the antigen which originally elicited that response, and secondly, the fact that certain dyes (fluorochromes) emit visible radiation after absorbing energy from short wavelength light or ultraviolet radiation. The first step in the technique is to obtain and purify antibody against the protein which is to be investigated. Next this antibody is conjugated with a fluorochrome. Then samples of cells or tissues are exposed to a solution of the labelled antibody, and excess antibody is removed. Finally, the preparation is examined via a fluorescence microscope.

Raising and Purifying the Antibody

A sufficient quantity of the protein, the presence and distribution of which is to be studied, must first be prepared in the purest possible form, in order to avoid the production of antibodies against other substances which may lead to invalid or confusing results. Rabbits are commonly used as the source of antibody, and the purified protein (antigen) is usually mixed with Freund's adjuvant and injected intramuscularly or subcutaneously. There are two forms of this adjuvant, known as 'incomplete' and 'complete'; the former consists of an emulsion of mineral oil and detergent with the antigen in aqueous solution, which promotes a stronger immune reaction by providing slow release of antigen over a period of several weeks. Complete Freund's adjuvant also contains killed mycobacteria, and evokes a chronic inflammatory lesion at the injection site, stimulating the animal's immune response even more strongly.

The aim of the initial injection is to establish a strong primary immune response, and a period of several weeks is required for this to be achieved. The animal can then be given one or more 'booster' injections of the antigen in saline to elicit a further increase in the amount of specific antibody in the blood, which may be taken about a week after the booster injection. The blood is allowed to clot, and the serum is obtained. The ultimate success of the technique depends very much on the presence of an adequate quantity of specific antibody in the serum. This may be assessed, for example, by a precipitin test in which the serum is mixed in tubes with a series of dilutions of the antigen; if there is sufficient antibody, a precipitate will quickly form along the whole length of the tube with one or more of the stronger antigen dilutions. The globulin fraction of the serum (which contains the antibody) is then extracted by mixing equal volumes of serum and a saturated solution of ammonium sulphate. This precipitates the globulins, which are then redissolved in saline and dialysed to remove the sulphate.

Monoclonal Antibodies

An antiserum prepared by the procedure outlined above suffers from some inherent disadvantages. These include the fact that it consists of a complex mixture of antibodies, each directed against a different antigenic determinant. Furthermore it is impossible to guarantee a continuing supply of antiserum of the same specificity as the original batch. These problems can now be avoided by using monoclonal antibodies, which are pure preparations of identical immunoglobulin molecules, specific for the same antigenic determinant. The principles on which the technique for producing monoclonal antibodies is based are as follows:

1. Each B-lymphocyte (or plasma cell) in an animal synthesises antibody specific for a single antigenic determinant. All the progeny of a particular B-lymphocyte will constitute a clone, every member of which will synthesise antibody against exactly the same determinant (i.e. a 'monoclonal' antibody). However, B-lymphocytes (or plasma cells) cannot be maintained indefinitely in culture.
2. There is a cancer of lymphoid tissue called a myeloma (plasma cell tumour), and cells from such a tumour can be isolated and *will* proliferate indefinitely in culture.
3. Some animal viruses (e.g. Sendai virus inactivated with ultraviolet light) can induce the fusion of single cells of different kinds, leading to the production of hybrid cells (Harris and Watkins, 1965).

It was Kohler and Milstein (1975) who first succeeded in preparing a monoclonal antibody. They had been using the cell fusion technique to produce hybrids from a mixture of mouse and rat myeloma cells. Milstein (1980) has described how he and his colleague Kohler conceived the idea that it might be possible to fuse a normal lymphocyte or plasma cell with a myeloma and thus obtain a hybrid which would proliferate indefinitely and whose descendents would all secrete antibody of exactly the same specificity. They chose sheep red blood cells as a test antigen because it is relatively easy to detect antibody against them. Having immunised mice against the sheep red cells, they collected cells from their spleens and mixed them with mouse myeloma cells in the presence of inactivated Sendai virus. The resulting fused spleen-myeloma hybrids were then cultured, and those which secreted antibody capable of causing lysis of sheep red cells were identified and individual clones were isolated (Kohler and Milstein, 1975).

When a sufficient number of cells of the appropriate clone have been obtained in this way they may be injected into animals of the same strain as those originally used to provide the cells which were fused. These host animals will then develop tumours (myelomas) secreting large quantities of the specific monoclonal antibody, which can be collected from their serum. Another method of producing the antibody in quantity is to grow large numbers of the cells in culture, and isolate the antibody from the culture medium. It is also possible to preserve a sample of the cells for future use by freezing them.

The availability of ample supplies of highly specific monoclonal antibodies against a wide variety of substances is already making, and will continue to make, a vital contribution in many areas of biological research, not least in immunocytochemistry.

Conjugation of the Antibody with a Fluorochrome

The two dyes which have been used more often than any others for immunofluorescence are fluorescein and rhodamine. The former has proved especially suitable because of its bright green fluorescence (easily distinguishable from blue autofluorescence) and the relative ease with which it can be coupled to antibody protein, but rhodamine, with its orange fluorescence, is a useful alternative, and has a number of advantages over fluorescein (McKay *et al.*, 1981). The fluorochrome must first be converted to a derivative containing reactive groups which will combine readily with protein. Fluorescein isothiocyanate (FITC) and tetramethylrhodamine isothiocyanate (TMRITC) are such derivatives, which are commercially available.

The process of conjugating fluorescein to antibody is, in essence, as follows. FITC is dissolved in buffer at pH 9.0 at a concentration of 1 mg/ml, and 1 ml of the solution per 100 mg of protein is added to the solution of antibody. The pH is adjusted to 9.5 while stirring, and the reaction is allowed to proceed for 60 minutes at room temperature. Unconjugated fluorochrome is then removed by means of column chromatography. Further purification may be necessary to eliminate unwanted labelled antibodies or other proteins so that the final solution contains, so far as possible, only fluorochrome-labelled antibody against the protein originally used as the antigen.

Figure 3.3: The direct immunofluorescence technique

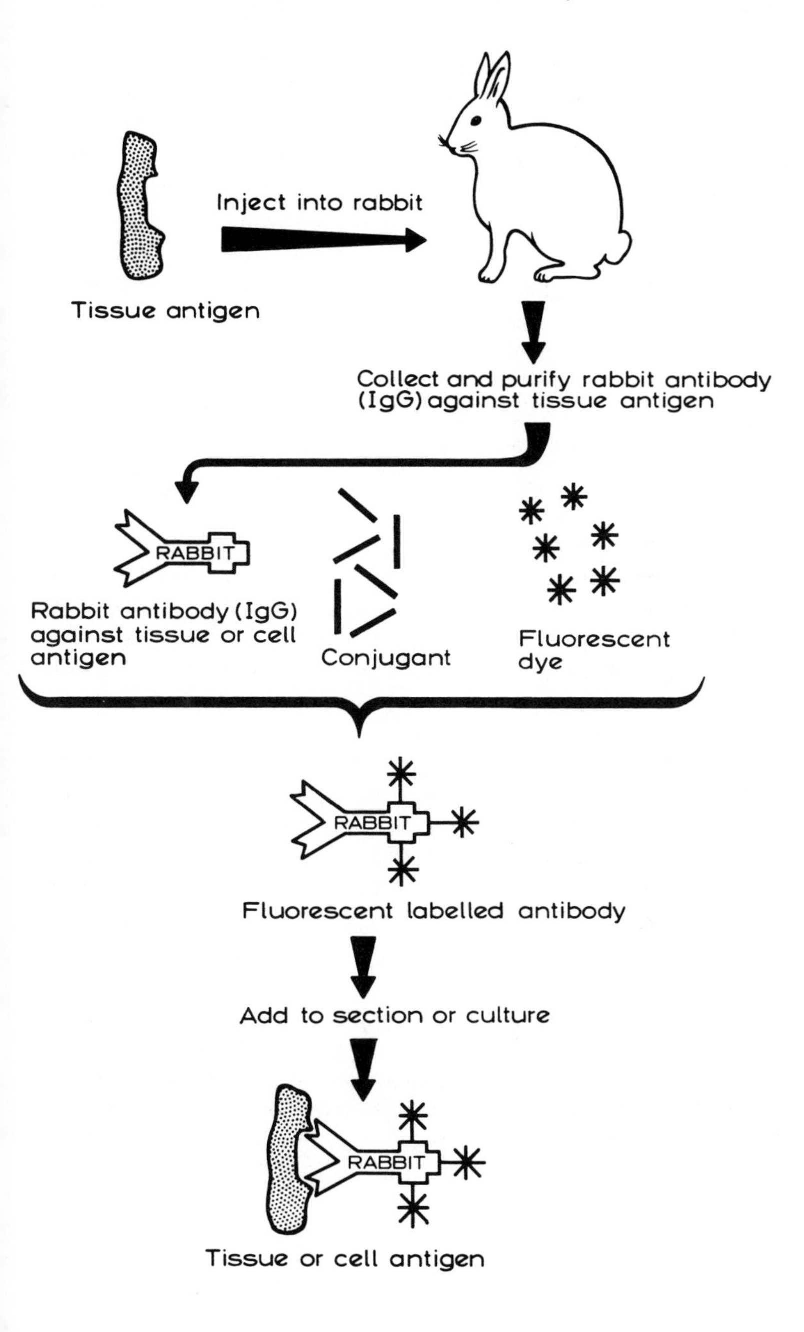

Staining Cells with Fluorescent Antibody

There are two ways of doing this, known as the 'direct' and the 'indirect' methods.

In the direct method, the antibody raised against the protein to be detected is conjugated with a fluorochrome, and the staining carried out simply by exposing cells to a solution of the labelled antibody (Figure 3.3).

The indirect method (sometimes known as the 'sandwich' technique) relies upon the fact that antibodies (immunoglobulins) are themselves capable of acting as antigens; thus, for example, if a goat is injected with rabbit immunoglobulins, goat antibody against rabbit immunoglobulins can be obtained and conjugated with a fluorochrome. The staining is then performed in two stages. First the cells are exposed to a solution of *unlabelled* rabbit antibody directed against the protein to be traced, and unbound antibody is removed. In the second stage, the sites where rabbit antibody has attached to the protein are labelled by exposure to goat fluorescent antibody against rabbit immunoglobulins (Figure 3.4).

The indirect method is more sensitive because several molecules of labelled anti-immunoglobulin antibody can react with each molecule of rabbit immunoglobulin, thus increasing the intensity of the fluorescence. A further advantage is that the same stock solution (such as goat fluorescent antibody against rabbit immunoglobulin) can be used to reveal the distribution of rabbit antibodies raised against any particular protein, and can be bought, already conjugated with fluorescein, from commercial sources.

Monolayers of cells (e.g. fibroblasts) are particularly suitable for the application of immunofluorescence methods, and the combination of the techniques of cell culture and immunocytochemistry has produced much new information about the distribution within cells of several proteins, such as actin, myosin and tubulin, which are directly (or indirectly) involved in cell locomotion. Antibody molecules will not penetrate the surface membrane of a living cell (though they may be actively taken into the cytoplasm in pinocytotic vesicles). Hence if intracytoplasmic structures are to be examined, it is necessary to make the cells permeable (e.g. with acetone) before they are covered with a solution of antibody. When the staining has been completed, the cells are given a final wash in phosphate-buffered saline and mounted in glycerol under a coverslip.

Figure 3.4: The indirect ('sandwich') immunofluorescence technique

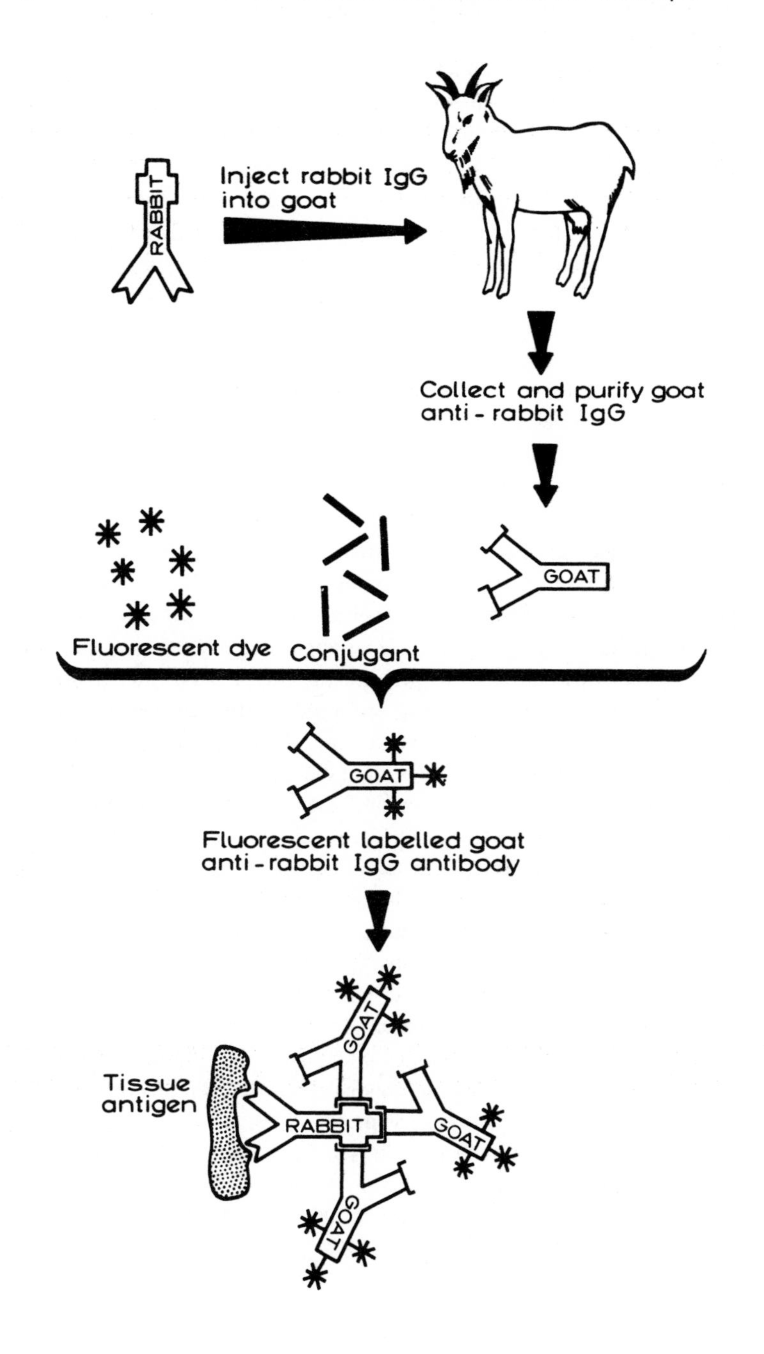

The Fluorescence Microscope

The fluorescence displayed by dyes such as fluorescein and rhodamine is an example of the phenomenon known as photoluminescence, and is subject to Stokes' law, which states that the wavelength of the emitted light is longer than (or equal to) that of the exciting light. Furthermore, the waves of visible light emitted by the fluorochrome will be smaller in amplitude than the exciting light. Account must also be taken of the fact that, for each fluorochrome, there is a narrow band of wavelengths of the exciting light which is maximally absorbed by the dye and generates maximal emission of fluorescence. For fluorescein, the optimum wavelength for excitation lies between 450 and 500 nm, whereas the emitted light is in the yellow-green region of the spectrum between 500 and 550 nm.

The application of these basic principles to the design of fluorescence microscopes has produced instruments with the following features:

1. A light source producing short wavelength ultraviolet and other excitation wavelengths at high intensity, in order to elicit as much fluorescence as possible from the fluorochrome. This requirement is best met by a 200 W mercury vapour lamp, which generates peaks of intense radiation in the short wavelength regions of the spectrum, including the optimum wavelengths for the excitation of fluorochromes.
2. As much as possible of the optimum exciting wavelengths should be allowed to reach the specimen, but other wavelengths should be prevented from doing so. This is achieved by placing an excitation (or primary) filter between the lamp and the specimen. Suitable filters are now available in the form of interference filters produced by depositing thin layers of metallic salts on glass; these will transmit a limited range of wavelengths but will reflect all other wavelengths. Thus when using fluorescein, an exciter filter which transmits wavelengths between 450 and 500 nm (the region of maximal absorption by the dye) will be selected.
3. As much as possible of the relatively faint specific fluorescence emitted from the dye should reach the observer's eye (or the photographic emulsion), but with as little as possible of the intense exciting radiation. This requirement is met by a barrier (or secondary) filter between the specimen and the eyepiece, which reflects all wavelengths except the band containing the emitted fluorescence (which will be of longer wavelength than the exciting radiation). For fluorescein, the barrier filter should transmit between 500 and 550 nm.

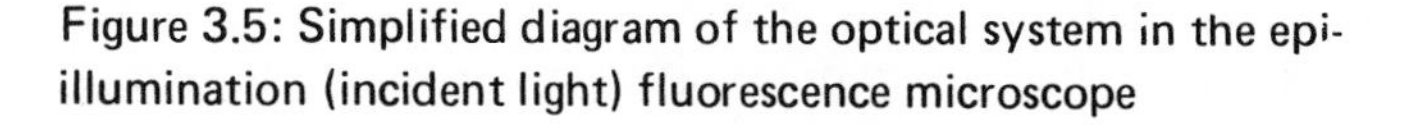

Figure 3.5: Simplified diagram of the optical system in the epi-illumination (incident light) fluorescence microscope

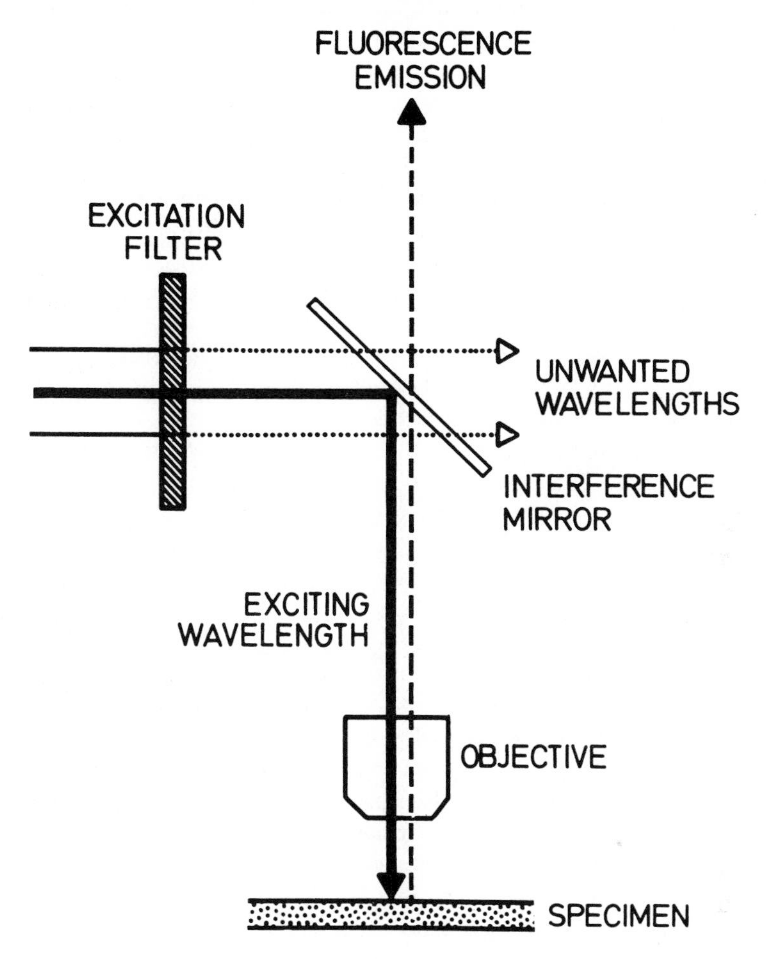

The early fluorescence microscopes were of the conventional type using transmitted illumination with an ordinary substage condenser, but some difficulty was experienced in preventing some excitation radiation from reaching the eye and drowning the fainter fluorescence. The use of a dark ground condenser went some way towards solving this problem, since the exciting light was not directed into the objective, but such condensers are not easy to operate. More recently, the tendency has been to use epi-illumination (incident light) for fluorescence microscopy. These instruments use an optical system very similar to that of the interference reflection microscope (see p. 30). The exciting radiation enters the microscope body horizontally, and is reflected down through the objective by an interference mirror (produced in the same way as the interference filters mentioned above). This radiation is focused onto the specimen by the objective, and light emitted by the fluorescing dye is collected by the objective and traverses the interference mirror to proceed up to the observer's eye (Figure 3.5).

Several advantages accrue from the use of epi-illumination:

1. The exciting illumination is automatically focused by the objective, and only the area of the specimen within the field of view is exposed to it, thus reducing the amount of fading which occurs.
2. The exciting radiation is directed away from the observer and cannot therefore drown the weaker fluorescence.
3. Fluorescence is excited in the uppermost part of the specimen, and the emitted fluorescence does not have to traverse the thickness of the specimen before entering the objective.

Electron Microscopy of Cultured Cells

In the early days of electron microscopy, before ultramicrotomes capable of cutting sections of the necessary thinness were available, the thinnest portions of cultured cells were used to produce some of the first electron micrographs of cytoplasmic organelles (Porter, Claude and Fullam, 1945; Porter, 1953). Hanging drop cultures of explants of chick embryo tissues were grown on coverslips previously coated with a thin layer of a polyvinyl resin (formvar). After 36-72 hours in culture the explants were discarded, and remaining cells in the outgrowth were fixed in osmium tetroxide. An area containing very thinly spread cells was selected, the formvar stripped from the glass and transferred to grids for insertion in the electron microscope. It was found that some areas of the cells were thin enough to be penetrated by electrons at 60 kV, and micrographs of mitochondria, Golgi apparatus and endoplasmic reticulum were obtained at magnifications up to × 15,000.

With the advent of microtomes capable of cutting ultra-thin plastic-embedded sections, whole cultured cells were superseded as material for electron microscopy, and attention was directed mainly to improving methods for the fixation, embedding, sectioning and staining of fragments of tissue excised from animals. However, the potential value of cultured cells as objects for electron microscopy was not lost sight of, and as routine electron microscope techniques were refined, they were applied to cells *in vitro*; indeed, transmission electron microscopy (TEM) of ultra-thin sections of cultured cells has been increasingly exploited in recent years by those interested in the cytoplasmic organelles which are directly implicated in cell locomotion. Even more recently, the examination of entire cultured cells has once again become a useful procedure both in the conventional TEM and in the high voltage electron microscope (HVEM). Scanning electron microscopy (SEM) has also been applied to cultured cells to produce detailed pictures of their surface configuration.

The Preparation of Cultured Cells for Ultra-thin Sectioning and Transmission Electron Microscopy

The procedures which have been developed for fragments of excised tissue may be applied to cells grown *in vitro*, with minor modifications; thus the fact that the cells are on a glass or plastic surface must be borne in mind during fixation, dehydration and embedding. In the case of cells grown on a glass coverslip, there may be difficulty in separating the polymerised embedding medium from the glass, although this can often be achieved by plunging the specimen into liquid nitrogen, when the glass may crack and separate, leaving the cells enclosed in the embedding medium. When the cells are grown on a plastic surface, substances which may dissolve the plastic must obviously not be used during the processing of the culture for electron microscopy.

The following schedule is fairly typical of the procedures generally used for the fixation and embedding of cultured cells for transmission electron microscopy:

1. Rinse the culture in phosphate-buffered saline;
2. Fix in buffered glutaraldehyde at 37°C for 10-20 min;
3. Rinse in buffer;
4. Post-fix in buffered osmium tetroxide 5-15 min;
5. Wash in distilled water three times;
6. Stain in 0.5-2% uranyl acetate, followed by a further wash in water;
7. Dehydrate in 35, 50, 70 and 90% ethyl alcohol (10 min in each);
8. Immerse in 90% hydroxypropyl methacrylate (HPMA), 3-5 changes over 15 min, then in 95% HPMA and 97% HPMA similarly;
9. Immerse in a mixture of HPMA and embedding resin (e.g. Epon),

then in 100% resin;
10. Drain off excess resin and polymerise.

When the thin sheet of polymerised resin containing the cells has been separated from the glass or plastic surface, it can be examined under a phase contrast microscope, and an area containing a particular cell (or group of cells) can then be marked and cut out, mounted on an ultramicrotome, and sectioned either at right angles to the original culture surface or parallel to it. Ultra-thin sections are then mounted on grids in the normal way for examination in the transmission electron microscope.

Transmission Electron Microscopy of Entire Cultured Cells

Interest in the possibility of obtaining high resolution and high magnification micrographs by passing the electron beam through the full thickness of a well-spread cultured cell, as originally attempted by Porter and his colleagues, has been increased by the development of the high voltage electron microscope (HVEM). In this instrument the accelerating voltage applied to the electron beam is one million or more, as compared with the 50-100 kV of the conventional electron microscope. The higher voltage gives the electrons greater penetration, and there is a slight increase in resolving power. In practice it has been found that, even in the 50-100 kV microscope, useful information can be acquired from the examination of entire cells (see, for example, Macartney and Parkinson, 1981). The following procedure is, therefore, applicable to the preparation of whole cells for examination either in the HVEM or in the conventional TEM, although with the latter only the thinnest parts of the periphery of the cells may be usable (Wolosewick and Porter, 1979).

Handling of the cells is much simplified by growing them from the outset on electron microscope grids; however, copper grids are potentially toxic to living cells, and gold or nylon grids are, therefore, preferable. The grids are covered with a film of formvar or celloidin, and placed on a coverslip; they are then coated with carbon, and sterilised in ultraviolet light. They are then placed in a Petri dish and immersed in a suspension of dissociated cells in nutritive medium. When the cells have settled on the grids in sufficient numbers (determined by examining the cultures with an inverted phase contrast microscope), they are fixed in glutaraldehyde and post-fixed in osmium tetroxide, using a schedule similar to the one outlined above.

Since the cells are to be examined intact and not sectioned on an ultramicrotome, it is not necessary to embed them, but they must be thoroughly dehydrated before being placed in the HVEM or the TEM. The fixed cells, still supported on the gold grids, are first dehydrated in ethanol-water mixtures with increasing concentrations of ethanol up to absolute alcohol. The final stage of dehydration is accomplished by the 'critical point' drying method. This involves placing the grids in a chamber filled with absolute ethanol (acetone or amylacetate may also be used). The ethanol is then replaced with liquid CO_2, and the temperature of the chamber is raised slowly to beyond the critical temperature for CO_2 (31°C). The liquid CO_2 is thus transformed to gas, which is allowed to escape slowly until the high pressure within the chamber has fallen to atmospheric pressure. The grids, bearing the now completely dried cells, are then removed, coated with carbon, and placed in the electron microscope.

The Preparation of Cultured Cells for Scanning Electron Microscopy

For this method of examination, it is particularly important to avoid exposing the living cells to anything which may distort or contaminate their surface. Thus it is advisable to give the cells a preliminary rinse in a balanced salt solution to wash away debris which may have settled onto them from the culture medium, but the saline must be warmed to 37°C and the cells must not be exposed to it for more than a few seconds before being immersed in 2-3 per cent buffered glutaraldehyde, also pre-warmed to 37°C. Fixation is then completed, again using a schedule similar to the one already outlined for the preparation of cultured cells for TEM. The coverslip on which the cells have been grown is then broken into fragments small enough to be attached to the SEM specimen holder. The cells are dehydrated in alcohol, and subjected to critical point drying, before the pieces of coverslip are mounted on specimen holders, and the cells are coated with carbon followed by gold or gold-palladium. They are then ready for examination in the SEM.

Immunocytochemistry and Electron Microscopy

The importance of fluorescent-labelled antibodies for studying the intracellular distribution of specific proteins with the light microscope is equalled (if not exceeded) by the potential value of antibodies labelled so that they are visible in the electron microscope.

In fact, an early method for the ultrastructural localisation of actin filaments did not involve the use of specific antibody, but relied upon the fact, first reported by Huxley (1963), that heavy meromyosin (HMM – the fragment of the complete myosin molecule which possesses ATP-ase activity and interacts with actin) will combine with isolated filaments of actin to produce a characteristic 'arrowhead' pattern. Ishikawa and his colleagues used this interaction to develop a technique for labelling actin filaments within cells (Ishikawa *et al.*, 1969). They treated cultures of skeletal muscle or chondroblasts, and fragments of tissue dissected from skin, intestine, trachea, spinal cord and gizzard taken from chick embryos, with glycerol, incubated them in a solution of HMM, and processed them for electron microscopy. The arrowhead pattern formed by the attachment of HMM to actin filaments was observed in muscle cells and in fibroblasts, chondroblasts, neurons and epithelial cells.

Antibodies Labelled with Ferritin or Horseradish Peroxidase

As in fluorescence immunocytochemistry, the specificity and purity of the antibody are of vital importance. When a suitable antibody has been prepared, there are two substances which are commonly used to render it visible in the electron microscope; these are ferritin and horseradish peroxidase.

Ferritin consists of a core of crystalline ferric hydroxide surrounded by an outer shell of protein. The core contains as many as 2,000 atoms of iron in a sphere only about 5 nm in diameter; hence its ability to scatter electrons. To attach ferritin to immunoglobulin requires the use of a coupling agent having two reactive groups, one to link to a molecule of ferritin, the other to a molecule of antibody. The reagent first used was xylene (or toluene)-2,4-diisocyanate (Singer and McLean, 1963). More recently, glutaraldehyde has been preferred as the coupling agent (Otto *et al.*, 1973). It has several advantages over the diisocyanate, being readily soluble in water and giving reproducible amounts of immunoferritin conjugates which can more easily be standardised. Glutaraldehyde was first used by Avrameas (1969) for conjugating antibody with enzymes in a one-stage process, but a two-stage method is preferable in order to give better control over the reaction and to reduce as much as possible the loss of specific antibody activity and the formation of unwanted aggregates. The principles underlying the conjugation of ferritin with immunoglobulin, and a recommended technique, will be found in Sternberger (1979). Staining with labelled antibody must be carried out before the cells (or the tissue specimen) are embedded, bearing in mind that the penetration of the antibody is retarded by cell surface membranes and by intracytoplasmic membranes. It can be increased by damaging these membranes, using a fixative (glutaraldehyde or formaldehyde) together with freezing and thawing, or organic solvents or detergents, although such treatment is likely to disturb the ultrastructure of the cells. Cells grown *in vitro* as a monolayer are, of course, more readily penetrated by the labelled antibody than cells within a solid fragment of tissue removed as a biopsy specimen.

The use of the enzyme, horseradish peroxidase, as a label was advocated by Nakane (Nakane and Pierce, 1966) and by Avrameas (Avrameas and Uriel, 1966) and their collaborators. The potential advantages of a suitable enzyme for this purpose include the ability of each molecule to catalyse a chemical change in several molecules of the substrate (thus increasing the local concentration of visible label) and the fact that it may be applicable to both light and electron microscopy. Horseradish peroxidase (HRP) was chosen because of its commercial availability and because of the existence of an established cytochemical technique for its localisation in electron micrographs (Graham and Kamovsky, 1966). Several coupling reagents were assessed for their suitability as a means of linking HRP to antibody, but it was found that glutaraldehyde gave the best results for both light and electron microscopy (Avrameas, 1969, 1970). Antibody labelled with HRP was employed originally, in much the same way as fluorescent-labelled antibody, using either a direct or an indirect method, but more recently a different technique, known as the 'Unlabelled Antibody Peroxidase-Antiperoxidase (PAP) Method', has been introduced. Devised by Mason *et al.* (1969) and by Sternberger and his colleagues (Sternberger, 1969; Sternberger *et al.*, 1970), this is designed to avoid or minimise some serious drawbacks inherent in the HRP-conjugated antibody method. During the chemical reaction involved in the conjugation process, significant damage is unavoidably done to the combining power of the antibody and to the activity of the enzyme used as the label. In the words of Sternberger (1979):

> The difficulties with labelled antibodies suggested that perhaps it might be worthwhile to attach a histochemically detectable marker to tissue antigen without the use of labelled antibodies at all. We felt that this could be easily accomplished if we were to use both combining sites of an antibody for two different purposes. We could use one of the combining sites for attachment to antigen in tissue, just as in labelled antibody techniques, and we could use the second

Figure 3.6: The unlabelled antibody peroxidase-antiperoxidase (PAP) technique, using goat anti-rabbit IgG

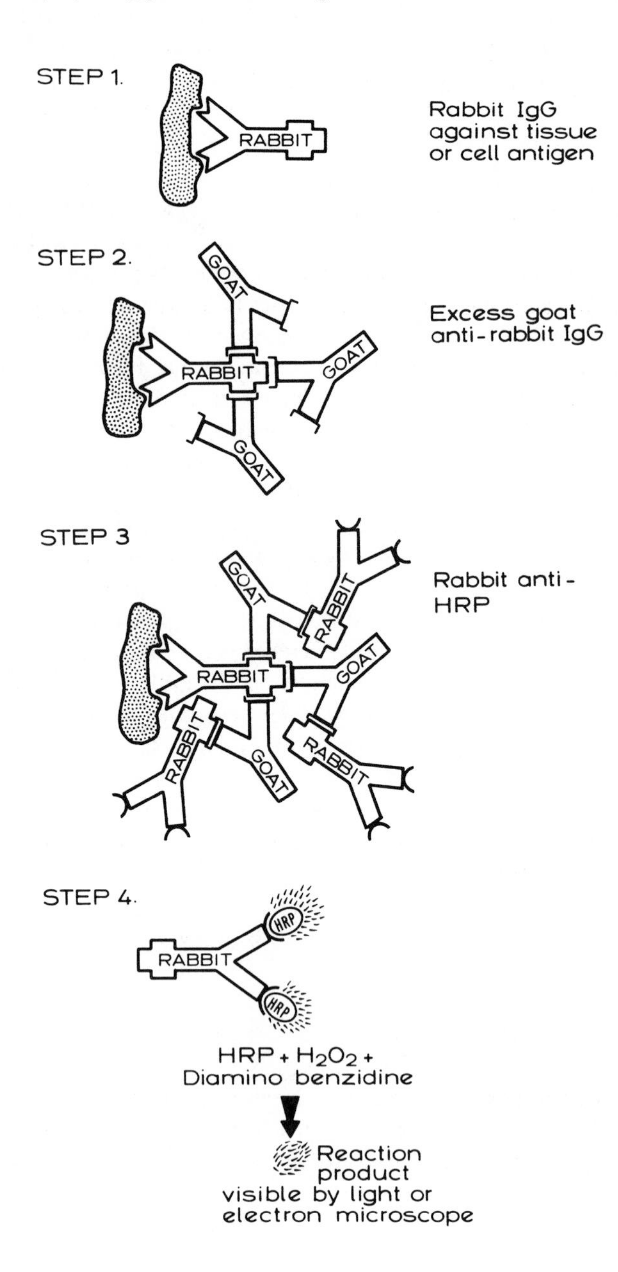

combining site to localize a histochemical marker to the antibody and thus, indirectly, to the tissue antigen site.

Sternberger and his colleagues assessed the practicability of the method, using a spirochaete, *Treponema pallidum*, as a test object. The procedure was carried out in five separate steps. First, spirachaetes on a slide were exposed to a solution of rabbit antibody against the organisms. Second, sheep antibody to rabbit immunoglobulin was used in excess, so that many of the sheep antibody molecules reacted with the rabbit immunoglobulin via only one of their two combining sites. Third, purified rabbit antibody against HRP was applied; this reacted with the remaining free combining sites of the sheep antibody used in step two. In the fourth step, HRP was added to react specifically with the combining sites of the rabbit anti-HRP antibody of step three. Finally, the HRP was allowed to act on hydrogen peroxide together with 3.3′-diaminobenzidine HCl, to produce an insoluble brown precipitate on the surface of the spirochaetes. The procedure is illustrated in Figure 3.6, using goat anti-rabbit IgG.

Later, it was found to be more convenient to telescope the third and fourth steps, so that instead of adding rabbit anti-HRP antibody and HRP separately, they were added simultaneously, in the form of a purified peroxidase-antiperoxidase (PAP) complex. The latter is composed chiefly of three molecules of HRP linked by two molecules of anti-HRP antibody (Figure 3.7).

Figure 3.7: Peroxidase-antiperoxidase (PAP) complex

As with ferritin conjugated antibodies, there are difficulties in applying the unlabelled antibody PAP technique because of poor penetration of tissues and cells; these problems are discussed by Sternberger (1979), who offers practical advice for their solution.

The advantages of the unlabelled antibody PAP method have been summarised by Mason *et al.* (1969):

> For example, fluorescent antibody methods necessitate specialized equipment. Precise identification of sites reacting with fluorescent antibody may be difficult because of the poor visualization of the nonfluorescing portions of the section with ultraviolet illumination. Comparison of sections stained by the fluorescent antibody and by reference methods is difficult and cannot at present be achieved directly in a comparison microscope as is possible with the enzyme-antibody methods. The transient nature of fluorescence may be a further disadvantage with the fluorescent antibody technique.
>
> Although the technique of conjugating enzymes to antibodies may circumvent the problems associated with the fluorescent method, the chemical conjugation required in this procedure has the possible handicap of affecting either the immunospecificity of the antibody or the activity of the conjugated enzyme. Moreover, antibody unconjugated during the coupling procedure could compete with conjugated antibody for antigenic sites, possibly reducing thereby the sensitivity of the procedure.
>
> The immunoglobulin-enzyme bridge technique avoids these problems associated with the fluorescent method and with the chemical conjugation procedure. The bridge method offers also the possibility of increased sensitivity through the multivalent reactivity of the immunoglobulins employed at the several steps in the procedure. Tissue stained for a specific antigen can be easily compared with serial sections stained for morphological reference, and brief osmication at the end of the enzymatic reaction produces a permanent slide preparation. In addition, the reagents used in the immunoglobulin-enzyme technique can be purchased commercially or can be easily made and are well preserved when frozen. Because the reaction product of the peroxidase substrate, diaminobenzidine, is electron opaque, the immunoglobulin-enzyme bridge may . . . be used to localize antigen with the electron microscope.

Antibodies Labelled with Colloidal Gold

The use of particles of gold as an electron dense label for antibodies relies on the fact that proteins can readily be absorbed onto them, and it has the advantage that, since no chemical reaction is needed to link the protein molecules to the gold, the antibody remains unchanged.

It was Faulk and Taylor (1971) who first capitalised on this; they prepared a suspension of gold particles from chloroauric acid, and coated them with rabbit anti-salmonella serum. Salmonella organisms were then exposed to the labelled antiserum, and examined in an electron microscope. Electron dense gold particles could be seen on the surfaces of the bacteria, separated from them by a gap which was presumed to be occupied by molecules of antibody.

When the technique was applied more extensively it was found that some immunoglobulins (including human and rabbit IgG) would not adhere firmly enough to the gold; Romano and Romano (1977) solved this problem by making use of a protein extracted from the cell wall of *Staphylococcus aureus*, known as staphylococcal protein A (SpA). This has a strong affinity for IgG molecules from several species of animals, and will also form a stable complex with colloidal gold. Romano and Romano exposed cells or viruses to specific antibodies against antigens on their surface, and then labelled the antibodies with gold particles coated with SpA.

At first, the SpA-gold method was used to identify antigens only on the external surfaces of cells, but Roth and his colleagues were able to extend its application to the detection of antigens within cells. This was achieved by incubating thin sections of cells, embedded in Epon, with a specific antiserum and then with SpA-gold complex (Roth *et al.*, 1978). By adding different amounts of sodium citrate to solutions of chloroauric acid, it is possible to produce suspensions of gold particles of different sizes. Bendayan (1982) took advantage of this to demonstrate two different antigens in the same plastic embedded section by applying different antisera to the two sides of the section, labelling one with SpA-gold particles averaging 12 nm and the other with SpA-gold particles averaging 19 nm in diameter.

References

S. Avrameas (1969) 'Coupling of Enzymes to Proteins with Glutaraldehyde', *Immunochemistry*, vol. 6, p. 43

— (1970) 'Immunoenzyme Techniques: Enzymes as Markers for the Localization of Antigens and Antibodies', *Int. Rev. Cytol.*, vol. 27, p. 349

— and J. Uriel (1966) 'Méthode de Marquage d'Antigènes et d'Anticorps avec des Enzymes et son Application en Immunodiffusion', *C.R. Acad. Sci. (D)* (Paris), vol. 262, p. 2543

M. Bendayan (1982) 'Double Immunocytochemical Labelling Applying the Protein A-Gold Technique', *J. Histochem. Cytochem.*, vol. 30, p. 81

A.H. Coons, H.J. Creech and R.N. Jones (1941) 'Immunological Properties of an Antibody Containing a Fluorescent Group', *Proc. Soc. exp. Biol.* (N.Y.), vol. 47, p. 200

—, H.J. Creech, R.N. Jones and E. Berliner (1942) 'The Demonstration of Pneumococcal Antigen in Tissues by the Use of Fluorescent Antibody', *J. Immunol.*, vol. 45, p. 159

A.S.G. Curtis (1964) 'The Mechanism of Adhesion of Cells to Glass', *J. Cell Biol.*, vol. 20, p. 199

W.P. Faulk and G.M. Taylor (1971) 'An Immunocolloid Method for the Electron Microscope', *Immunochemistry*, vol. 8, p. 1081

D. Gingell (1981) 'The Interpretation of Interference-Reflection Images of Spread Cells: Significant Contributions from Thin Peripheral Cytoplasm', *J. Cell Sci.*, vol. 49, p. 237

R.C. Graham and M.J. Kamovsky (1966) 'The Early Stages of Absorption of Injected Horseradish Peroxidase in the Proximal Tubules of Mouse Kidney: Ultrastructural Cytochemistry by a New Technique', *J. Histochem. Cytochem.*, vol. 14, p. 291

H. Harris and J.F. Watkins (1965) 'Hybrid Cells from Mouse and Man: Artificial Heterokaryons of Mammalian Cells from Different Species', *Nature*, vol. 205, p. 640

H.E. Huxley (1963) 'Electron Microscope Studies on the Structure of Natural and Synthetic Filaments from Striated Muscle', *J. Mol. Biol.*, vol. 7, p. 281

H. Ishikawa, R. Bischoff and H. Holtzer (1969) 'Formation of Arrowhead Complexes with Heavy Meromyosin in a Variety of Cell Types', *J. Cell Biol.*, vol. 43, p. 312

C.S. Izzard and L.R. Lochner (1976) 'Cell-to-Substrate Contacts in Living Fibroblasts: an Interference Reflexion Study with an Evaluation of the Technique', *J. Cell Sci.*, vol. 21, p. 129

G. Kohler and C. Milstein (1975) 'Continuous Cultures of Fused Cells Secreting Antibody of Predefined Specificity', *Nature*, vol. 256, p. 495

J.C. Macartney and E.K. Parkinson (1981) 'The Ultrastructure of Whole Cells from Two Human Keratinocyte Strains During Cell Spreading and Island Formation *in vitro*', *Exp. Cell Res.*, vol. 132, p. 411

I.C. McKay, D. Forman and R.G. White (1981) 'A Comparison of Fluorescein Isothiocyanate and Lissamine Rhodamine (RB200) as Labels for Antibody in the Fluorescent Antibody Technique', *Immunology*, vol. 43, p. 591

T.E. Mason, R.F. Phifer, S.S. Spicer, R.A. Swallow and R.B. Dreskin (1969) 'An Immunoglobulin-Enzyme Bridge Method for Localizing Tissue Antigens', *J. Histochem. Cytochem.*, vol. 17, p. 563

C.A. Middleton (1973) 'The Control of Epithelial Cell Locomotion in Tissue Culture' in *Locomotion of Tissue Cells*, Ciba Foundation Symposium 14 (Elsevier, Amsterdam), pp. 251-62

C. Milstein (1980) 'Monoclonal Antibodies', *Scientific American*, vol. 243, p. 66

P.K. Nakane and G.B. Pierce (1966) 'Enzyme-Labelled Antibodies: Preparation and Application for the Localization of Antigens', *J. Histochem. Cytochem.*, vol. 14, p. 929

J.F. Nunn, J.A. Sharp and K.L. Kimball (1970) 'Reversible Effect of an Inhalational Anaesthetic on Lymphocyte Motility', *Nature* vol. 226, p. 85

H. Otto, H. Takamiya and A. Vogt (1973) 'A Two Stage Method for Cross-Linking Antibody Globulin to Ferritin by Glutaraldehyde. Comparison Between the One-Stage and the Two-Stage Method', *J. Immunol. Meth.*, vol. 3, p. 137

J. Padawer (1968) 'The Nomarski Interference-Contrast Microscope. An Experimental Basis for Image Interpretation', *J. roy. Mic. Soc.*, vol. 88, p. 305

K.R. Porter (1953) 'Observations on a Submicroscopic Basophilic Component of Cytoplasm', *J. exp. Med.*, vol. 97, p. 727

—, A. Claude and E.F. Fullam (1945) 'A Study of Tissue Culture Cells by Electron Microscopy', *J. exp. Med.*, vol. 81, p. 233

O.W. Richards (1954) 'Phase Microscopy 1950-1954', *Science*, vol. 120, p. 631

E.L. Romano and M. Romano (1977) 'Staphylococcal Protein A Bound to Colloidal Gold: A Useful Reagent to Label Antigen-Antibody Sites in Electron Microscopy', *Immunochemistry*, vol. 14, p. 711

J. Roth, M. Bendayan and L. Orci (1978) 'Ultrastructural Localization of Intracellular Antigens by the use of Protein A-Gold Complex', *J. Histochem. Cytochem.*, vol. 26, p. 1074

S.J. Singer and J.D. McLean (1963) 'Ferritin-Antibody Conjugates as Stains for Electron Microscopy', *Lab. Invest.*, vol. 12, p. 1002

L.A. Sternberger (1969) 'Some New Developments in Immunocytochemistry', *Mikroskopie*, vol. 25, p. 346

—. P.H. Hardy, J.J. Cuculis and H.G. Myer (1970) 'The Unlabelled Antibody Enzyme Method of Immunohistochemistry. Preparation and Properties of Soluble Antigen-Antibody Complex (HRP-AntiHRP) and Its Use in Identification of Spirochaetes', *J. Histochem. Cytochem.*, vol. 18, p. 315

Suggestions for Further Reading

U. Gröschel-Stewart (1980) 'Immunochemistry of Cytoplasmic Contractile Proteins', *Int. Rev. Cytol.*, vol. 65, p. 193

R.C. Nairn (1976) *Fluorescent Protein Tracing*, 4th edn (Churchill Livingstone, Edinburgh)

A.G.E. Pearse (1980) *Histochemistry, Theoretical and Applied*, 4th edn (3 vols, Churchill Livingstone, London), vol. 1, chap. 6, 'Immunocytochemistry', pp. 159-252

P.N. Riddle (1979) *Time-lapse Cinemicroscopy*, 1st edn (Academic Press, London)

K.F.A. Ross (1967) *Phase Contrast and Interference Microscopy for Cell Biologists*, 1st edn (Edward Arnold, London)

L.A. Sternberger (1979) *Immunocytochemistry*, 2nd edn (John Wiley and Sons, New York)

J.J. Wolosewick and K.R. Porter (1979) 'Preparation of Cultured Cells for Electron Microscopy', in K. Maramorosch (ed.), *Practical Tissue Culture Applications* (Academic Press, New York), pp. 59-85

4 CYTOPLASMIC FILAMENTS

When cells are treated with non-ionic detergents, under appropriate conditions, most of the proteins are extracted but a framework of fibrous components remains. This framework is known as the 'cytoskeleton', and it consists of microfilaments, intermediate filaments and microtubules. It is now known that the cytoskeleton is intimately involved in producing, co-ordinating and directing both movement of the cell as a whole (i.e. cell locomotion) and the movement of various components within the cell. Hence we shall, in this and the following chapter, discuss what is presently known of the structure and functions of the various elements of the cytoskeleton.

It was perhaps inevitable that muscle, and especially striated voluntary muscle with its obvious involvement in the movements of the body and its highly ordered structure, should first attract the attention of scientists interested in the analysis of the cellular basis of movement. By the 1950s it was known that two proteins, actin and myosin, are primarily responsible for muscle contraction. Later, due to the work of H.E. Huxley and his colleagues, it became clear that muscle contraction occurs when filaments of actin and myosin slide past each other, and that the mechanism requires ATP and is regulated by the presence or absence of calcium ions.

Although it was well known that many kinds of cell, other than muscle fibres, were capable of displaying movements, knowledge of the cytoplasmic structures responsible for such movements in nonmuscle cells continued to be scanty for some time after the elucidation of the contractile mechanism in muscle. As late as 1964, Wohlfarth-Botterman, in a review article, listed the following as the currently most popular theories which had been propounded to explain movements in amoebae:

1. Active shearing forces (parallel lengthwise shifting of cytoplasmic fibrils).
2. Diffusion-drag force hypothesis (pull or drag due to transport of matter caused by diffusion).
3. Potential difference theory (the presence of a potential difference in the cytoplasm of an amoeba along the axis of its motion might control its movement).
4. Sol-gel transformation theories (particularly popular, based on

transformations between ectoplasm and endoplasm).
5. Contractility (first suggested by F.E. Schulze in 1875).

A propos the last theory, Wohlfarth-Botterman (1964) wrote:

> There is no doubt that the cytoplasm of amebae is basically capable of contracting. It is an open question, however, which structures are able, through active contraction, to furnish motive force for ameboid movement, nor are the details of the mechanism of motion, or how it is controlled, fully understood.

Although we are not yet able to give a complete answer to this question, the fact that we are now able to provide at least a partial answer is the result of work carried out by many scientists during the last 30 years. An experiment of particular significance was done by A.G. Loewy (1952); he prepared extracts from the cytoplasm of the slime mould *Physarum polycephalum* and was able to show that the viscosity of these extracts was altered by the addition of ATP in a similar way to the change in viscosity seen in mixtures of actin and myosin extracted from muscle. Loewy wrote:

> The basic postulate pervading this work is the assumption that what is true for muscle may also be true for less specialized cells or, to put it into evolutionary phraseology, the highly specialized mechanism in muscle must find its origin in the primitive cell.

Later, Bettex-Galland and Luscher (1959) were able to prepare an actomyosin-like extract from human blood platelets. These results clearly suggested that contractile proteins were not restricted to muscle cells. A similar conclusion was inferred from experiments with glycerinated cells. Treatment of muscle fibres with glycerol removes soluble proteins, leaving a cytoskeleton consisting mainly of actin and myosin. The addition of Mg^{2+} and ATP to such preparations causes them to contract like intact muscle fibres. Hoffmann-Berling and Weber (1953) pioneered the use of similar techniques on nonmuscle cells and were able to show that glycerinated amoebae and fibroblasts would both contract when treated with Mg^{2+} and ATP. It was, however, not until 1966, when Hatano and Oosawa purified actin from the cytoplasm of *Physarum polycephalum*, that actin was unequivocally proved to be present in nonmuscle cells. Shortly afterwards, Hatano and Tazawa (1968) were able to extract myosin from the same source, and subsequently both actin and myosin have been found in a wide variety of cells from both plants and animals.

Actin

Up to 15 per cent of the total protein in actively motile cells consists of actin. The actins extracted from different sources differ slightly from each other in their amino-acid sequence but are all very similar to muscle actin (suggesting that actin has been highly conserved during evolution).

All the actins are proteins composed of about 375 amino-acid residues, with a molecular weight *circa* 42,000 daltons; hence, they all migrate together during electrophoresis on sodium dodecyl sulphate/polyacrylamide gels. The differences detected between them take the form of small changes in the primary structure of the molecules and differences in their isoelectric points, but their significance (if any) is unknown.

Figure 4.1: Diagram of the F-actin helix

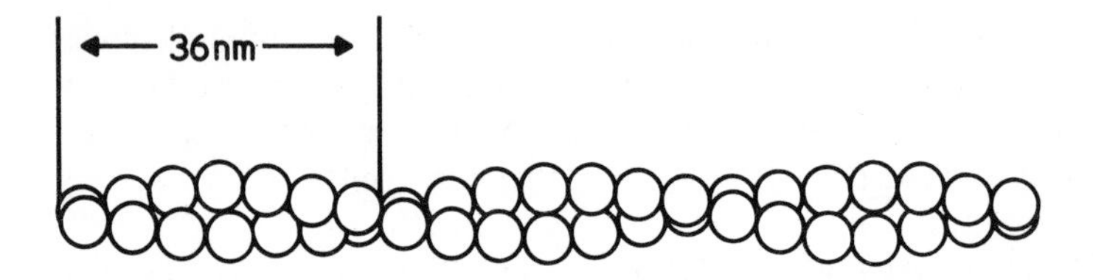

In solution, actin takes the form of a globular subunit with a diameter of about 5.5 nm. If the salt concentration in the solution is increased, these monomers of globular actin (called G-actin) polymerise spontaneously to form filamentous or F-actin. The filaments consist of two chains of subunits wound around each other in a helix approximately 7.0 nm in diameter. There are about 13 G-actin monomers in each repeat unit of the helix, giving the filaments an axial repeat of about 36 nm (Figure 4.1). This substructure gives the actin filaments a characteristic beaded appearance when viewed in the electron microscope after negative staining.

F-actin forms the backbone of the thin filaments of striated muscle. In nonmuscle cells much of the actin is also found in the form of cytoplasmic filaments of F-actin (see p. 64), but there is evidence to suggest that in some types of cell a proportion of the actin exists in an unpolymerised state. It has been estimated that as much as 50 per cent of the actin from chick brain cells and tissue cultured fibroblasts may be present in the unpolymerised form (Bray and Thomas, 1976). This may represent a pool of stored actin which can be converted into filaments when required.

Myosin

Cytoplasmic myosin accounts for about 0.5-1.5 per cent of the total protein in nonmuscle cells and is therefore present in much smaller amounts than actin. The chemical and physical properties of myosins from different sources differ considerably, but, in general, most of the cytoplasmic myosins from nonmuscle cells show considerable similarities to the myosin of smooth muscle fibres (but have less in common with myosin from striated muscle). Most nonmuscle myosins have a molecular weight of around 470,000 daltons and are composed of two heavy chains (MW about 200,000 daltons) and two pairs of light chains (MW about 17,000 and 20,000 daltons, respectively). The molecule has a globular head region composed of part of the heavy chains and all of the light chains, and a helical rod or tail region composed of the remainder of the heavy chains (Figure 4.2). A noteworthy exception to these generalisations is the myosin isolated from *Acanthamoeba castellanii*, which has a molecular weight of about 180,000 daltons and is made up of one heavy (MW about 140,000) and two light chains (MW about 16,000 and 14,000, respectively).

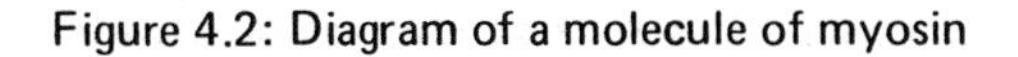
Figure 4.2: Diagram of a molecule of myosin

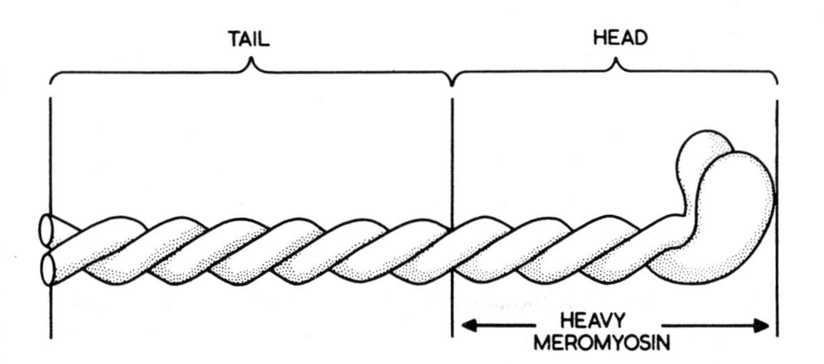

Under physiological conditions myosin molecules aggregate spontaneously to form bipolar filaments, consisting of a bare central shaft made up of the tails of the molecules, with their globular heads projecting at each end; such aggregates form the thick filaments in striated muscle. In nonmuscle cells it has proved difficult to identify myosin filaments unambiguously, and the localisation of myosin in these cells is still a matter of debate.

Associated Proteins

There are other proteins, in addition to actin and myosin, which are thought to contribute, directly or indirectly, to contractile mechanisms. Some contribute by regulating the interaction between actin and myosin, others may maintain actin and myosin in an appropriate spatial relationship.

In vertebrate striated muscle the interaction between actin and myosin is regulated by two proteins, tropomyosin and troponin. In brief, a complex, consisting of tropomyosin and three subunits of troponin, binds to actin and, in the absence of Ca^{2+}, prevents its interaction with myosin. In the presence of Ca^{2+} the affinity between one of the subunits of troponin and tropomyosin is reduced; this causes the exposure of reactive sites on the actin molecule which then interacts with myosin.

Tropomyosin has been isolated from a number of nonmuscle cells, including cells from calf brain and pancreas, and from cultured mouse fibroblasts (see Gröschel-Stewart, 1980). Immunofluorescence has shown that it may be found associated with actin filaments (Lazarides, 1975). Troponin has not yet been demonstrated convincingly in nonmuscle cells, but there are preliminary reports of the occurrence of proteins with troponin-like properties in such cells (Gröschel-Stewart, 1980).

An example of a protein which may be involved in maintaining actin and myosin in the correct relationship is α-actinin, which is known to be localised in the Z-lines of striated muscle and may link the actin filaments between sarcomeres; α-actinin has been identified, by both biochemical and immunofluorescence techniques, in a number of nonmuscle cell types (e.g. Lazarides and Burridge, 1975).

A number of proteins capable of modifying the state of actin in nonmuscle cells have been identified. Some of these form cross links between actin filaments and promote their gelation, whilst others inhibit the polymerisation of actin and may also induce the disassembly of actin filaments (Schliwa, 1981). An example of the first type is filamin, an actin binding protein first isolated from smooth muscle and subsequently identified in nonmuscle cells (Wang and Singer, 1977).

Microfilaments

The presence of fibres within the cytoplasm of cultured cells has been recognised at least since the description by Lewis and Lewis (1924) of 'tension striae', which they saw in cultures of mouse endothelium and mesothelium. Because these striae are frequently found in regions where the cytoplasm appears to be under stress or tension, they are often called 'stress fibres', and are found in many different types of cell *in vitro*. Their presence in living cells can be detected using bright field, phase contrast, Nomarski interference contrast, or polarised light microscopy. In fixed cells they can also be detected by electron microscopy and immunofluorescence.

With phase contrast optics, stress fibres in living fibroblasts appear as dark lines, often orientated parallel to the long axis of the cell (Figure 4.3). In polarised light they show positive birefringence with respect to their long axes. With the advent of the electron microscope it became clear that stress fibres consist of bundles of parallel closely packed filaments 5-7 nm in diameter (Buckley and Porter, 1967), now referred to as microfilaments. Subsequent investigations have shown that microfilament bundles (stress fibres) are often associated with contacts between the plasma membrane of the cell and the substratum. In such areas the bundles seem to be attached to the plasma membrane and to run obliquely back into the cytoplasm, rising gradually from the lower towards the upper surface of the cell and usually ending in a region of fibrous material around the nucleus (Figure 4.4). The electron microscope also reveals that microfilaments are not confined to these bundles; filopodia contain a single bundle of parallel microfilaments, and microfilaments are also present in a cortical layer of cytoplasm immediately beneath the plasma membrane on both upper and lower surfaces of the cell (Figure 4.4). This cortical layer of microfilaments is of variable thickness (0.1-0.5 μm) and, in contrast to the orderly arrangement of the filaments in microfilament bundles, it contains an irregular meshwork of microfilaments. On the lower surface of the cell the cortical layer is interrupted by the obliquely running stress fibres (Figure 4.4).

As we shall see later (p. 105), bundles and meshworks are interchangeable forms of microfilament organisation.

Figure 4.3: Phase contrast photomicrograph of a living chick embryo heart fibroblast showing numerous stress fibres

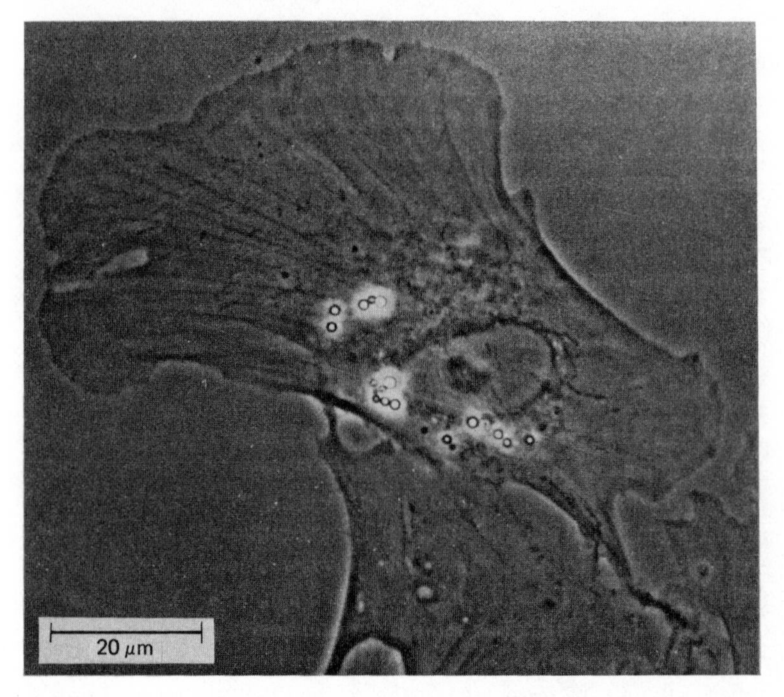

Figure 4.4: Diagram of a vertical longitudinal section of a fibroblast in culture showing the distribution of microfilaments. The cortical meshwork extends into surface protrusions called lamellipodia (see p.112). Filopodia (see p. 61) contain a single bundle of microfilaments. (Modified from Willingham *et al.*, 1981.)

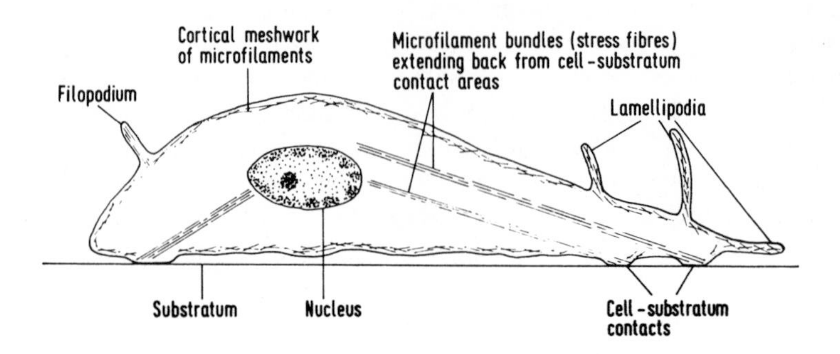

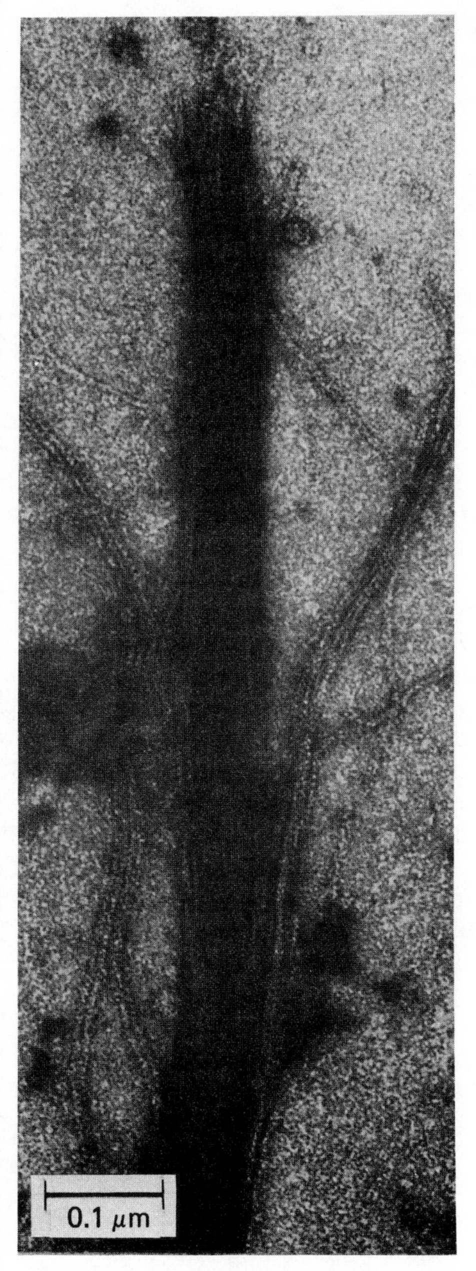

Figure 4.5: Electron micrograph of a negatively stained bundle of microfilaments from the green alga *Nitella flexilis.* (From B.A. Palevitz, J.F. Ash and P.K. Helper, *Proc. Nat. Acad. Sci.*, vol. 71 (1974), pp. 363-6.)

Composition of Microfilaments and Microfilament Bundles

The ultrastructural similarities between the microfilaments of non-muscle cells and the thin filaments of muscle encouraged speculation that microfilaments might themselves consist of actin. That this is the case was established by the heavy meromyosin decorating technique, described in Chapter 3 (see p. 48), and applied by Ishikawa *et al.* (1969) to cultured cells. The specific 'arrowhead' pattern of distribution of HMM along the microfilament bundles in stress fibres and filopodia, and on individual microfilaments in the cortical meshwork, confirmed that actin is the major component in these cytoplasmic structures in nonmuscle cells.

A combination of HMM decorating, negative staining and electron microscopy has shown that microfilaments have a double helical structure with a diameter of 40-70 nm. They have a characteristic beaded appearance with an axial repeat of about 37 nm and thus have a structure very similar to the thin (actin) filaments of muscle (Figure 4.5).

Immunofluorescence is particularly valuable as a means of revealing the intracellular distribution of actin in microfilaments. When cultured cells are fixed, rendered permeable with acetone, and incubated with anti-actin antibody, brightly fluorescent fibres can be seen within the cells using either direct or indirect fluorescence techniques (see Chapter 3, p. 40). There is a precise correspondence between the distribution of these fluorescent fibres and that of microfilament bundles (stress fibres) seen by phase contrast optics (Figure 4.6). Ultrastructural observations using the indirect immunoperoxidase technique (see Chapter 3, p. 49) have confirmed that the anti-actin antibody is localised in the microfilament bundles.

Immunofluorescence techniques, of course, lack the resolving power to reveal the precise distribution of actin in individual microfilaments in the cortical meshwork, and this region of the cytoplasm merely displays diffuse fluorescence when these techniques are used.

Immunofluorescence has shown that microfilament bundles also contain other proteins thought to be implicated in mechanisms of contraction; these include proteins antigenically similar to myosin, tropomyosin, α-actinin and filamin (reviewed by Gröschel-Stewart, 1980).

Figure 4.6: A culture of chick embryo heart fibroblasts, (a) phase contrast, (b) the same cells photographed via the fluorescence microscope after the application of the indirect immunofluorescence method using anti-actin antibodies. (Courtesy of Dr J.P. Heath.)

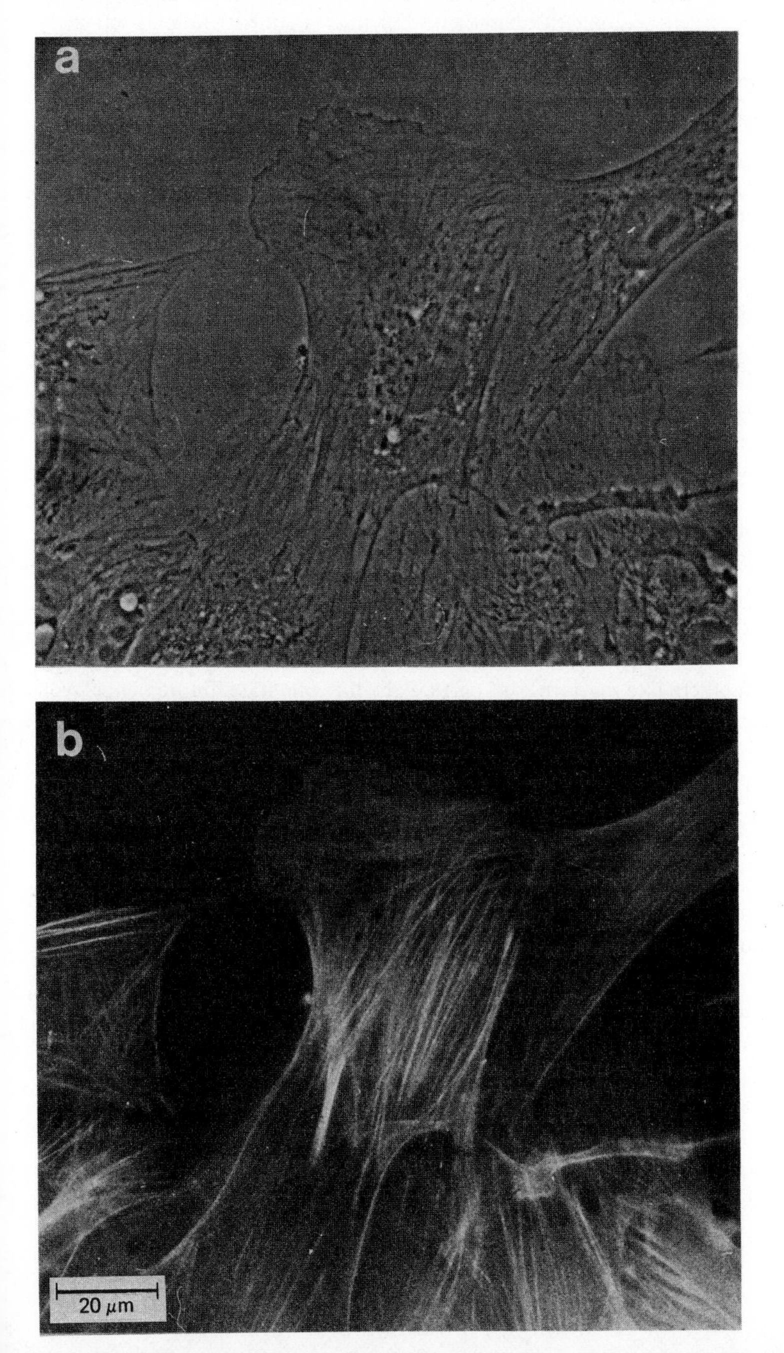

When anti-actin antibody is used, microfilament bundles are seen to fluoresce fairly uniformly from end to end (Figure 4.6), but this is not the case with antibodies against some of the other proteins. Thus anti-myosin antibodies often reveal a discontinuous or punctate pattern of fluorescence along the length of the microfilament bundles, strongly fluorescent regions alternating with non-fluorescing regions (in the BHK cell line the former are roughly 0.6 μm, the latter 0.47 μm, in length – but precise measurements are difficult to make) (Goldman *et al.*, 1979).

When antibodies against α-actinin or tropomyosin are used, a discontinuous pattern of fluorescence is also seen along microfilament bundles in human and rat fibroblasts; with anti-α-actinin the segments of fluorescence are about 0.25-0.50 μm in length, separated by non-fluorescing segments about 1.0-1.4 μm long. Anti-tropomyosin antibody creates a pattern of fluorescence complementary to that produced by anti-α-actinin, with the fluorescent segments approximately 1.2 μm long and the non-fluorescent about 0.4 μm in length (Lazarides, 1975; Lazarides and Burridge, 1975). Thus it seems that α-actinin and tropomyosin may alternate along the microfilament bundles, a possibility which is supported by the finding that sequential staining with antibodies against these two proteins reveals microfilament bundles that fluoresce uniformly throughout their length (Lazarides and Burridge, 1975).

The discontinuous distribution of myosin, tropomyosin and α-actinin revealed by these experiments suggests that the three proteins may be highly organised within microfilament bundles, perhaps in a sarcomere-like pattern. In particular, the punctate fluorescence seen with anti-α-actinin antibody is (at least superficially) similar to the fluorescent banding seen when striated muscle fibres are treated with this antibody. This banding is known to be caused by the localisation of α-actinin in the Z-lines of muscle. It is possible, therefore, that within microfilament bundles of nonmuscle cells, Z-line-like segments containing α-actinin alternate with segments containing tropomyosin (and possibly myosin). However, caution is necessary when deducing molecular organisation from images produced by immunofluorescence techniques; the binding of antibodies to a structure exaggerates its size, since a molecule of IgG, for example, has a diameter of about 9 nm, and this problem is exacerbated when more than one layer of antibody is present (as in the indirect technique). Further uncertainty arises from the possibility that steric hindrance may make some antigenic sites inaccessible to the antibody, and procedures used to render the cell more permeable to antibodies may remove some of the proteins from it, or alter the arrangement of those that remain.

Fortunately, there is evidence, derived from a different technique, suggesting that the molecular organisation inferred from immunofluorescence may be correct. This evidence has been obtained by injecting fluorescent-labelled specific proteins directly into living cells and monitoring their subsequent distribution within the cytoplasm. In this way, actin labelled with rhodamine shortly after injection into a living cell *in vitro*, is found to be uniformly distributed throughout the cytoplasm, but 30 minutes later much of it has become localised in microfilament bundles, which then fluoresce with equal intensity along their whole length (Kreis, Winterhalter and Birchmeier, 1979). Using rhodamine labelled α-actinin, a similar experiment has shown that this protein becomes localised in microfilament bundles with a punctate distribution (Kreis and Birchmeier, 1980). These observations, then, support the possibility that the distribution of these proteins in the living cell is similar to that inferred from the results of immunofluorescent methods applied to fixed cells.

These observations have been extended by the use of antibodies labelled with ferritin so that they can be located in electron micrographs (see Chapter 3, p. 48). Herman and Pollard (1981) have used ferritin labelled anti-myosin antibodies to study the localisation of myosin in cultured HeLa cells. In microfilament bundles, the electron microscope revealed patches heavily labelled with ferritin, alternating with less heavily labelled regions. If confirmed, this result would strongly support the presence of alternate areas of high and low myosin concentration along the microfilament bundles giving, perhaps, a more accurate picture of the distribution of myosin than that provided by immunofluorescence, which seemed to suggest an absence of myosin in the non-fluorescing regions. We should therefore be cautious in placing too much reliance on the accuracy of the apparent distribution of α-actinin and tropomyosin revealed by immunofluorescence methods.

Ultrastructural observations have also revealed the possibility of an ordered arrangement of the components within microfilament bundles; with conventional fixation, alternate regions of higher and lower electron density have been reported in the bundles in some cells. If tannic acid is added to the glutaraldehyde during fixation, this pattern is accentuated, and microfilament bundles from a number of sources show electron dense regions (dark bands) with a fairly regular spacing. Their average length varies considerably in different types of cell, from 0.4 μm in gerbil fibroma cells (Gordon, 1978) to as little as 0.16 μm in PtK_2 cells (Sanger and Sanger, 1980). The significance of this banding pattern is unknown; Sanger and Sanger (1980) have suggested that in PtK_2 cells the less electron dense regions contain myosin and the bands of greater density α-actinin, and on this basis have proposed a sarcomere-like model for the arrangement of the various proteins in the bundles of microfilaments. But Goldman *et al.* (1979) have proposed that, in BHK21 cells, it is the dense bands which contain myosin, and this is supported by the observation that, in HeLa cells, the electron dense regions show the heaviest labelling with ferritin conjugated anti-myosin antibody (Herman and Pollard, 1981). Since this antibody produces significant but less dense labelling of the lighter bands (see above), it is premature to conclude that myosin is exclusively localised in the dark bands.

Both immunofluorescence and ultrastructural studies suggest the possibility that the various proteins in microfilament bundles may be arranged in a highly ordered fashion, but a clear understanding of their spatial interrelationships is not possible without further investigation at the ultrastructural level. More use of antibodies conjugated with electron dense markers is likely to be particularly useful. Until we have greater knowledge of the form and arrangement of these proteins in microfilament bundles, models of the bundles will be no more than speculative. It seems clear, however, that microfilament bundles are capable of contraction, at least in model systems. Using laser beam microdissection, Isenberg *et al.* (1976) isolated individual microfilament bundles from fibroblasts derived from a rat mammary adenocarcinoma (a remarkable feat in view of the fact that they are only about 1 μm in diameter), and were able to show that under appropriate conditions they would contract in the presence of ATP. Similarly, using video intensification, Kreis and Birchmeier (1980) have shown that, in glycerinated fibroblasts, microfilament bundles stained with rhodamine-labelled α-actinin contract under the influence of ATP. As yet, though, similar contraction has not been observed in living cells.

The state of knowledge about the organisation of myosin, tropomyosin, α-actinin and filamin in the microfilament network in the cortical region of the cytoplasm (see p. 61) is even more unsatisfactory. Immunofluorescence with antibodies against these proteins shows that they are present (e.g. Gottlieb *et al.*, 1979; Lazarides, 1976) but provides no clue about their precise distribution. It seems likely that techniques with greater resolution than can be provided by the fluorescence microscope will be needed; a preliminary study using anti-myosin antibody conjugated to ferritin does suggest that, in HeLa cells, there is a concentration of myosin in regions of the cytoplasm *not* associated with microfilament bundles (Herman and Pollard, 1981). It is interesting that much of the ferritin labelling in the general cytoplasm – and that in the microfilament bundles (see above) – was not found to be associated with recognisable myosin filaments. Perhaps local conditions somehow prevent the spontaneous assembly of myosin filaments?

An alternative approach has been to isolate microfilaments and to characterise their component molecules. In this way Schloss and Goldman (1980) have shown (as expected) that microfilaments extracted from nonmuscle cells contain actin, and have obtained evidence that it is associated with tropomyosin. The further development of this and other techniques will be essential to give us a fuller picture of the localisation and organisation of the proteins in microfilaments of the cortical meshwork. However, what immunofluorescence has done is to reveal something of the association between microfilaments and α-actinin and tropomyosin, which occurs during the aggregation of individual microfilaments into bundles.

Cytoplasmic Extracts

Interactions between contractile proteins have continued to be investigated in cytoplasmic extracts (see p. 57). Allen *et al.* (1960) showed that when cytoplasm extracted from *Chaos chaos* was isolated within capillary tubes, it retained the ability to stream and contract; Thompson and Wolpert (1963) demonstrated that cytoplasmic extracts of *Amoeba proteus* would gel and contract when warmed to room temperature in the presence of ATP. More recently, considerable impetus has been given to this approach by Kane's work on extracts of sea urchin eggs. If these extracts are prepared at 4°C and then warmed to 35-40°C with the addition of ATP, they gel spontaneously and then contract (Kane, 1975). These gels are packed with filaments of F-actin, and Kane (1975; 1976) has produced evidence that the sol-gel transition involves the polymerisation of G-actin to F-actin and requires the presence of two other proteins with molecular weights of 58,000 and 220,000 daltons, which apparently interact with the actin filaments and cross-link them. Extracts with similar properties have also been prepared from rabbit macrophages (Hartwig and Stossel, 1975). If ice cold extracts from this source are warmed to room temperature in the presence of ATP, they gel and then slowly contract. Using purified actin, Stossel and Hartwig (1976) found that the only other component needed to induce the formation of a gel is an actin-binding protein (ABP) with a molecular weight of 220,000 daltons, and that the gel will contract reversibly if myosin is added. Similar behaviour has been observed in cytoplasmic extracts from the slime mould *Dictyostelium discoideum* (Condeelis and Taylor, 1977).

The results of these various experiments suggest that actin gels may have a dual function in the cytoplasm of living cells. Firstly, actin gels formed by the cross-linking of microfilaments could act as part of the 'cytoskeleton' giving mechanical support to the cell cortex and related structures such as membrane ruffles and filopodia. Secondly, cell motility could be achieved by myosin-induced reversible contractions of the gels.

The situation is complicated by the fact that a number of diverse proteins appear to be capable of inducing the formation of an actin gel, or of modifying the extent of gel formation (Schliwa, 1981; Weeds, 1982). Furthermore, at least two proteins (profilin and actin inhibitor) are known to be capable of preventing the polymerisation of G-actin to F-actin, and are evidently responsible for keeping much of the cellular actin in an unpolymerised state (see p. 59). The role of these various proteins in the living cell is unknown, but it seems likely that some of them will be found to play an important part in regulating the state of cellular actin and thus influencing the shape and movement of the cell.

The Cytochalasins

The name of this group of fungal metabolites is derived from the Greek, cytos = a cell, chalasis = relaxation; they inhibit (reversibly) a number of cellular activities, including locomotion *in vitro*, cytokinesis, blood clot retraction, the contraction of smooth muscle, and cytoplasmic streaming (see Tanenbaum, 1978, for a review). In 1971, Wessells *et al.* drew attention to the fact that, in many cases, the inhibition of some aspect of cell motility by cytochalasin B was accompanied by the disruption of the cell's microfilaments; on withdrawing the drug the microfilaments reappeared and the motility was resumed. They therefore suggested that cytochalasin B inhibited cellular contractile processes by interfering directly with the microfilaments, and predicted that 'sensitivity to the drug implies (the) presence of some type of contractile microfilament system'. This was based on circumstantial evidence, and subsequent efforts to demonstrate a direct effect of the cytochalasins on microfilaments and actin produced conflicting and confusing results. This uncertainty was increased when it was discovered that cytochalasins inhibited sugar transport into cells. Since this effect was almost certainly due to interference with a function of the cell surface membrane, it was possible that the effects on cell motility might result from the same cause, rather than from a direct effect on microfilaments within the cytoplasm. The situation was clarified to some extent when a derivative of cytochalasin B, dihydrocytochalasin B, was synthesised and proved to have little or no influence on sugar transport but inhibited cell movements, indicating that the drug interfered with these two activities by independent mechanisms (Lin *et al.*, 1978). Since then it has become clear that cytochalasins B, D and E interact directly with purified actin filaments in such a way that they preferentially reduce the rate of addition of G-actin monomers to one end of the filaments (Brown and Spudich, 1979; Lin *et al.*, 1980). One method of demonstrating this has been to decorate filaments of F-actin with HMM and then to monitor the rate at which undecorated G-actin monomers are added to the filaments. Under these conditions monomers are added to both ends of the filaments but with a ten-fold bias in favour of addition to the barbed end of the decorated filament. In the presence of 2μM cytochalasin B the total rate of polymerisation is reduced by approximately 90 per cent, but the addition of monomer to the 'pointed' end of the filaments is little changed (Maclean-Fletcher and Pollard, 1980). Hence it seems likely that cytochalasin B interferes with the addition of monomers mainly at the 'barbed' end of the actin filaments. The stoichiometry of the binding of cytochalasin to F-actin strongly suggests that there are very few binding sites on each filament. Indeed, Brown and Spudich (1981) have shown, by using actin filaments of various lengths, that the number of binding sites for cytochalasin B is proportional to the number of filament ends and is independent of the concentration of actin, indicating that the binding sites are located only at the ends of the actin filaments and, presumably, at the 'barbed' ends.

It is known that the cytochalasins inhibit gel formation in cytoplasmic extracts (or reduce their viscosity); this has been established with extracts from pulmonary macrophages (Hartwig and Stossel, 1979), HeLa cells (Weihing, 1976), and *Acanthamoeba* (Pollard, 1976). What is still unclear is the mechanism by which cytochalasin produces this effect. Some observations suggest that cytochalasin B can cause a reduction in the length of purified actin filaments, and this might be expected to inhibit gelation induced by cross-linking agents (Hartwig and Stossel, 1979). However, the drug can also inhibit the gelation of actin extracts even when cross-linking agents are not required and this has led to the suggestion that the cytochalasins may inhibit some direct association between actin filaments necessary for gel formation (Maclean-Fletcher and Pollard, 1980).

There is further uncertainty about the way in which the cytochalasins influence living cells. After exposure to the drug, cytoplasmic microfilaments are usually described as being 'disrupted' or 'disorganised'. In a combined immunofluorescence and electron microscopic study, Weber *et al.* (1976) found that, in mammary fibroblasts, carcinoma cells and 3T3 cells, cytochalasin B caused the microfilament bundles to disappear; at the same time, star-shaped 'heaps' of aggregated actin and tropomyosin were formed, and were seen in electron-micrographs to consist of dense masses of condensed microfilamentous materials. A tentative explanation of this effect is that the disruption of microfilament bundles results from the depolymerisation of F-actin to G-actin. Alternatively, since cytochalasin inhibits actin polymerisation, it may interfere with a persisting cycle of actin polymerisation and depolymerisation which may normally occur in the cytoplasm. In either case, treatment of living cells with cytochalasin would be expected to cause a net increase in the proportion of G-actin within them. The evidence for this remains equivocal. Cytochalasin D causes morphological changes in cells of the human tumour line HEp-2 but does not seem to increase the proportion of G-actin in them (Morris and Tannenbaum, 1980), so there may be no net depolymerisation of F-actin. But work on the activation of human platelets by thrombin points to the opposite conclusion; in unstimulated platelets about 35-50 per cent of the actin is in a filamentous form and this rises to about 70 per cent after stimulation by thrombin. Cytochalasin D inhibits this effect, presumably by preventing the polymerisation of G-actin (Fox and Phillips, 1981; Casella *et al.*, 1981). In this system cytochalasin D actually seems to bring about the active depolymerisation of F-actin, since treatment of platelets already stimulated by thrombin causes an increase in G-actin level to that found in unstimulated platelets (Casella *et al.*, 1981). There is some indication that, at least in platelets, there may be a pool of stabilised F-actin which is insensitive to cytochalasin, since platelets unstimulated by thrombin do not show any increase in G-actin following treatment with cytochalasin D (Casella *et al.*, 1981).

The ability of cytochalasins to inhibit gel formation in cytoplasmic extracts, or to reduce their viscosity, may also be relevant to their known effects on living cells. If, as seems likely, actin networks contribute to the cytoskeleton, then the cytochalasins could conceivably operate by interfering with the filament-filament or filament-crosslinking protein interactions which seem to be important for the preservation of actin networks and gels. Such interference would presumably disorganise and destabilise the cytoplasm and might result in the impairment of cellular movements. In a recent study, Schliwa (1982) has confirmed that cytochalasin D disrupts the ordered cytoskeletal network of cultured BSC-1 cells. However, he suggests that this results from the drug breaking actin filaments rather than inhibiting filament interactions. This primary effect of the drug is independent of energy metabolism but is followed by a secondary, energy dependent, response. This results in the formation of dense aggregates containing actin filaments, myosin and tropomyosin, apparently as a result of uncontrolled and disorganised cytoplasmic contractions (Schliwa, 1982).

At present, therefore, it remains possible that cytochalasins may operate via two mechanisms, one affecting the polymerisation and depolymerisation of actin, the other interfering with interactions between actin filaments.

Intermediate Filaments

In addition to microfilaments, another apparently important filamentous system is found in the cytoplasm of many cells from higher eukaryotes. This system, first described in detail by Ishikawa *et al.* (1968) in cultured skeletal muscle cells, consists of filaments with a diameter of about 10 nm; because they are wider than the 6 nm actin filaments, but narrower than the 15 nm myosin filaments, they are known collectively as intermediate filaments (Ishikawa *et al.*, 1968). Since 1968, intermediate filaments with diameters of 8-12 nm have been demonstrated in a wide variety of cell types, both *in vivo* and *in vitro*. In the electron microscope the intermediate filaments from different cells look very similar, but biochemical and immunological techniques have revealed that it is possible to distinguish several different classes of these filaments. On the basis of the fact that some types of filament are found only in certain tissues, and taking account of their subunit composition, it was proposed that five distinct classes could be defined (Lazarides, 1980):

(i) Keratin (or prekeratin) filaments restricted to epithelial cells.
(ii) Neurofilaments, found characteristically in neurons.
(iii) Glial filaments, in glial cells.
(iv) Desmin filaments, found mainly in smooth, skeletal and cardiac muscle.
(v) Vimentin filaments, found in cells of mesenchymal origin such as fibroblasts.

It is now evident that the distribution of desmin and vimentin is not so restricted as was originally thought, since these proteins occur, singly or together, in many different cell types. It seems that desmin and vimentin filaments may be two members of the same class of intermediate filaments (Lazarides, 1981) so that there may, in fact, be only four distinct classes of these filaments. However, such a classification can only be tentative at present.

All the classes of intermediate filaments have other significant features in common, in addition to their width. All are relatively insoluble at physiological ionic strengths and pH, but under appropriate conditions they can all be dissociated into proteinaceous subunits which can then be repolymerised *in vitro* (see Lazarides, 1981; Anderton, 1981). Further, all the intermediate filament subunits so far identified are α-helical α-type fibrous proteins which apparently exist in a coiled-coil configuration (Steinert *et al.*, 1980).

Intermediate filaments of the keratin class are found in most epithelial cells, not merely in those which actually undergo keratinisation (Sun *et al.*, 1979). They consist of several different keratin polypeptides (MW 42,000-65,000 daltons) and are apparently identical with the tonofilaments seen in electronmicrographs of epithelial cells (Franke *et al.*, 1978a). Immunofluorescence observations on epithelial cells in culture reveal arrays of 6-11 nm thick keratin filaments, often organised into wavy bundles which are particularly abundant around the nucleus, forming a perinuclear 'cage'. The bundles radiate from the nuclear region into the thinner peripheral cytoplasm, and often end, apparently, by inserting into desmosomes (Osborn *et al.*, 1978).

Neurofilaments *circa* 10 nm in diameter are characteristically found in neurons. In vertebrates they consist of three major polypeptides with MWs of 210,000, 160,000 and 68,000 daltons. Invertebrate neurofilaments apparently consist of only two major polypeptides; those from the giant axon of the marine fanworm *Myxicola infundibulus* have molecular weights of 160,000 and 150,000 daltons, while those from the squid giant axon have molecular weights of 200,000 and 60,000 daltons (see Anderton, 1981, for a review). Immunofluorescence with antibodies against neurofilament proteins, both *in vivo* and *in vitro*, has confirmed that these filaments are restricted to neurons and has shown that they are not present in glial cells (Liem *et al.*, 1978; Yen and Fields, 1981). In neurons *in vitro* the neurofilaments form a poorly defined meshwork in the cytoplasm of the perikaryon, but they stain more intensely in the cell processes (Yen and Fields, 1981).

Glial filaments about 8 nm in diameter are found in astrocytes and other glial cells, but are absent from neurons. They consist of a single major polypeptide with a molecular weight of 51,000 daltons known as glial fibrillary acidic protein (GFAP) (see Lazarides, 1980, for a review). The localisation of GFAP in glial filaments has been confirmed by immunofluorescence and immunocytochemical techniques (Liem *et al.*, 1978; Steig *et al.*, 1980; Yen and Fields, 1981). Immunofluorescent staining of cultured glial cells using anti-GFAP antibody reveals sweeping bundles of glial filaments in the perinuclear region, which extend into the cytoplasmic processes (Antanitus *et al.*, 1975; Yen and Fields, 1981).

Intermediate filaments occur in all three types of muscle, but are most abundant in adult smooth muscle. The major subunit of these filaments is a protein with a molecular weight of *circa* 50,000 daltons which has been named 'desmin'. In smooth muscle cells the filaments form a network which links cytoplasmic dense bodies to membrane-bound dense plaques (see Lazarides, 1980, for a review). Immunofluorescence has shown that, in embryonic cardiac muscle and skeletal muscle myotubes in culture, desmin forms an intricate filamentous network. However, in adult cardiac and skeletal muscle, desmin is found in the Z-discs of the myofibrils. When the Z disc is seen *en face* in sections at right angles to the length of the myofibril, desmin is found at the periphery of the disc where it forms a transverse network encircling and interlinking adjacent Z discs (see Lazarides, 1980, for a review). There is therefore a redistribution of desmin from free cytoplasmic filaments to the Z disc during the differentiation of both cardiac and skeletal muscle (Gard and Lazarides, 1980).

Fibroblasts, and other cells of mesenchymal origin, contain a network of *circa* 10 nm intermediate filaments, the major subunit of which is a protein, with a molecular weight of 52,000 daltons, named vimentin. At first, vimentin was thought to be restricted to cells of this type but more recently vimentin filaments have been detected in a wide variety of different types of cell (Franke *et al.*, 1979a). Immunofluorescence of cultured cells shows that characteristically wavy vimentin fibres extend through the cytoplasm in a more-or-less radial fashion, and are especially abundant around the nucleus (Franke *et al.*, 1978b).

A number of cell types are known to have intermediate filament subunits of more than one type; thus several varieties of epithelial cell in culture have been shown by immunofluorescence, and by biochemical techniques, to contain both keratin and vimentin (Franke *et al.*, 1979b). Smooth muscle, skeletal muscle and fibroblasts have similarly been shown to contain both desmin and vimentin (Gard *et al.*, 1979), and glial cells to contain both GFAP and vimentin (Yen and Fields, 1981). In the case of epithelial cells, keratin and vimentin can be seen in immunofluorescent preparations to be present in separate arrays of intermediate filaments (Franke *et al.*, 1979b), but in muscle cells the distribution of desmin coincides with that of vimentin (Gard and Lazarides, 1980). However, it has not been established whether the two proteins copolymerise into the same filaments or whether there are two separate sets of filaments with identical distributions in the cell. The coincident distribution of desmin and vimentin in muscle cells, together with similarities in their peptide maps and the widespread occurrence of one or both of the proteins in a variety of cell types, suggests that they may be related members of a single class of intermediate filament subunit present in most cells (see Lazarides, 1981, for a review). Recently, a third possible member of this class has been found; this protein, known as 'synemin', has a molecular weight of 230,000 daltons and copurifies through cycles of depolymerisation and repolymerisation with desmin and vimentin from smooth muscle. Immunofluorescence in skeletal muscle reveals that synemin coincides in its distribution with desmin and vimentin throughout myogenesis (Granger and Lazarides, 1980), and in avian erythrocytes synemin also has a coincident distribution with vimentin (Granger *et al.*, 1982). It has therefore been proposed that vimentin, desmin and synemin may all occur as subunits of the same filament, and that in different cells the function of the filament may be altered by changing the proportions of the three subunits (Lazarides, 1981).

The fact that similar physical properties are shared by all intermediate filament subunits has prompted the suggestion that they may all contain domains with homologous amino-acid sequences. Such a constant domain could explain the ability, which they have in common, to polymerise into filaments about 10 nm in diameter. But the fact that many varieties of cell have more than one type of subunit suggests that each subunit may have a unique function, with the implication that, in addition to the constant domain, each subunit may have a variable domain responsible for its specific function (Lazarides, 1980). Some support for the concept of a constant domain has emerged from partial amino-acid sequence data which show considerable sequence identity between cleavage fragments from vimentin, desmin and the 68,000 dalton neurofilament subunit (Geisler *et al.*, 1982), but confirmation must await the complete sequencing of all the intermediate filament subunits.

The function of the intermediate filaments is obscure; in skeletal muscle desmin may link transversely individual myofibrils at their Z discs and link the discs to the plasma membrane, thus integrating mechanically the contractile actions of the muscle fibre. Desmin may also be involved in the attachment of actin to membranes in other muscle and nonmuscle cells (see Lazarides, 1980). Keratin filaments are often found inserted into desmosomes, and may therefore play a role in the mechanical support of epithelial cell layers (Drochmans, *et al.*, 1978). Similarly, neurofilaments may form a structural lattice providing tensile strength in axons (Gilbert *et al.*, 1975). Accumulations of vimentin filaments are often found in the perinuclear area, where they could give mechanical support to the nucleus and maintain its position within the cell (Small and Celis, 1978). In the light of these suggestions, Lazarides (1980) has proposed that intermediate filaments are involved in the mechanical integration of the various components of the cytoplasmic space. If this is the case, intermediate filaments might be expected to be capable of interactions with many other cytoplasmic structures. In addition to the evidence that they are capable of interacting with desmosomes, membranes and nuclei, there is some reason to think that they may be capable of interacting with microtubules; in many cells *in vitro*, disruption of the microtubules with drugs such as colchicine and vinblastine induces the formation of perinuclear whorls of neuro-, glial, desmin and vimentin filaments (Wisniewski *et al.*, 1968; Yen and Fields, 1981; Franke *et al.*, 1979b; Gard and Lazarides, 1980). Thus the arrangement of the intermediate filaments seems to be influenced by the state of microtubule organisation. The distribution of keratin filaments is relatively unaffected by such treatments (Franke *et al.*, 1979b).

Further progress in understanding the functions of intermediate filaments is hampered by the unavailability, at present, of drugs or procedures which specifically affect them. One approach has been to microinject cells with monoclonal antibody against intermediate filaments. When 3T3 cells are so treated, the network of vimentin filaments collapses into a perinuclear bundle without affecting the distribution of the microtubules. Although this particular experiment produced no detectable change in the behaviour, shape or movement of the injected cells (Klymkowsky, 1981), similar attempts to correlate alterations in the organisation of intermediate filaments with changes in cellular activities may well improve our knowledge of their function.

The Microtrabecular Lattice

Ultrastructural observations on cells prepared for electron microscopy by conventional techniques usually reveal the various cellular components, including those of the cytoskeleton, suspended in apparently amorphous cytoplasm known as the ground substance. There is some evidence, however, to suggest that the ground substance may not be as structureless as it seems in such preparations. Much of this evidence derives from the use of the high voltage transmission electron microscope (HVTEM, see p. 46), which allows images to be obtained from relatively thick specimens. Thin cytoplasmic extensions of cultured cells can thus be examined after fixation and critical point drying (see p. 47) without the need for sectioning. In cultured fibroblasts prepared in this way the ground substance appears as an irregular three-dimensional lattice of slender strands extending throughout the cytoplasm. Since this structure is reminiscent of the trabecular organisation of spongy bone, it has been named the 'microtrabecular lattice' (Wolosewick and Porter, 1976). A similar lattice structure has been seen in the ground substance of a number of other cell types. The individual microtrabeculae are usually 3-6 nm in diameter and 100 nm or more in length; between them are intertrabecular spaces or channels 50-150 nm in width (Wolosewick and Porter, 1979). The microtrabecular lattice is apparently continuous with the cortical layer of cytoplasm beneath the plasma membrane and its strands appear to support various organelles. In particular, the endoplasmic reticulum, polysomes, microfilaments and microtubules are coated with lattice material continuous with that of the microtrabeculae, and it may be that they are suspended within the lattice (Wolosewick and Porter, 1976). On the basis of these observations, Wolosewick and Porter (1979) have proposed the model of microtrabecular organisation shown in Figure 4.7. Further, they have suggested that the lattice divides the interior of the cell into two phases; a polymerised protein-rich phase constituting the lattice itself and a fluid water-rich phase filling the intertrabecular spaces. They consider that the microtrabecular lattice may be involved in organising and co-ordinating the activities of the organelles suspended within it and may thus play an important part both in regulating and directing transport within the cell and in controlling the shape and movement of the cell as a whole (Wolosewick and Porter, 1976).

However, the reality of the microtubular lattice is disputed, and it is considered by some investigators to be merely an artefact resulting from changes in the cytoplasmic proteins induced by some of the techniques used in the preparation of cells for electron microscopy. Small (1981) has produced evidence that the network of actin filaments in the leading edge of cultured fibroblasts can be converted to an irregular fibrous lattice, reminiscent of microtrabeculae, by certain dehydration techniques and also by osmication and critical point drying. Similarly, Heuser and Kirschner (1980) have found that in freeze-dried fibroblasts from mice the cytoplasm of the leading edge consists of fibrous cytoskeletal elements embedded in a granular matrix, but that prefixation with glutaraldehyde can convert this arrangement to one resembling a microtrabecular lattice. The view that the microtrabecular lattice is an artefact has to some extent been countered by the demonstration that a microtrabecular lattice is present in a variety of cells prepared for electron microscopy by a number of different procedures, and that the formation of a similar lattice cannot be induced in model systems of protein solutions by such procedures (Wolosewick and Porter, 1979). It would be premature, though, to conclude that the microtubular lattice is a genuine cytoplasmic component superimposed on the microfilaments, intermediate filaments and microtubules of the cytoskeleton.

Figure 4.7: Diagram illustrating the microtrabecular lattice. (Redrawn from Wolosewick and Porter, 1979.)

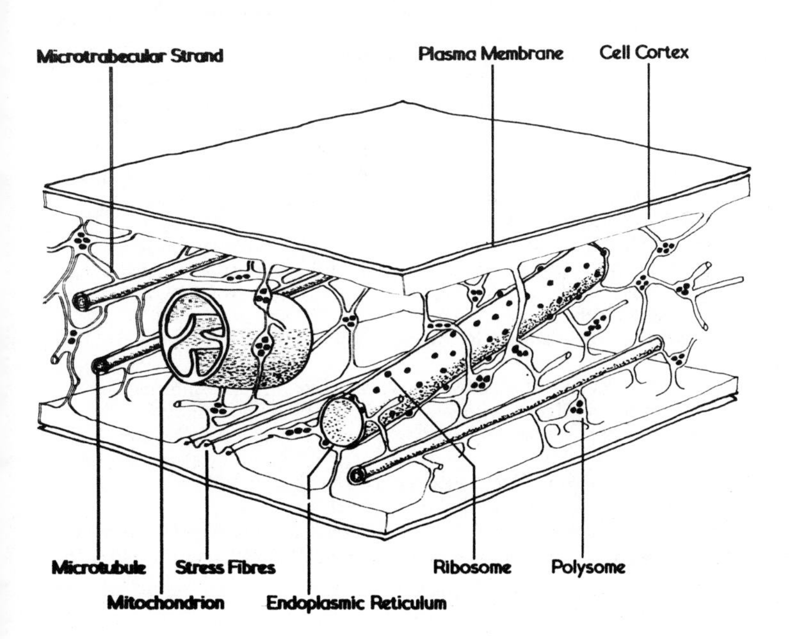

References

R.D. Allen, J.S. Cooledge and P.J. Hall (1960) 'Streaming in Cytoplasm Dissociated from the Giant Amoeba, *Chaos chaos*', *Nature*, vol. 187, p. 896

B.H. Anderton (1981) 'Intermediate Filaments: a Family of Homologous Structures', *J. Muscle Res. Cell Motility*, vol. 2, p. 141

D.S. Antanitus, B.H. Choi and L.W. Lapham (1975) 'Immunofluorescence Staining of Astrocytes *in vitro* Using Antiserum to Glial Fibrillary Acidic Protein', *Brain Res.*, vol. 89, p. 363

M. Bettex-Galland and M. Luscher (1959) 'Extraction of an Actomyosin-like Protein from Human Thrombocytes', *Nature*, vol. 184, p. 276

D. Bray and C. Thomas (1976) 'Unpolymerized Actin in Tissue Cells', in R. Goldman, T. Pollard and J.L. Rosenbaum (eds.), *Cell Motility* Book A (Cold Spring Harbor), pp. 461-73

S.S. Brown and J.A. Spudich (1979) 'Cytochalasin Inhibits the Rate of Elongation of Actin Filament Fragments', *J. Cell Biol.*, vol. 83, p. 657

— and J.A. Spudich (1981) 'Mechanism of Action of Cytochalasin: Evidence That it Binds to Actin Filament Ends', *J. Cell Biol.*, vol. 88, p. 487

I. Buckley and K. Porter (1967) 'Cytoplasmic Fibrils in Living Cultured Cells. A Light and Electron Microscope Study', *Protoplasma*, vol. 64, p. 349

J.F. Casella, M.D. Flanagan and S. Lin (1981) 'Cytochalasin D Inhibits Actin Polymerization and Induces Depolymerization of Actin Filaments Formed During Platelet Shape Change', *Nature*, vol. 293, p. 302

J.S. Condeelis and D.L. Taylor (1977) 'The Contractile Basis of Amoeboid Movement. V. Control of Gelation, Solation and Contraction in Extracts from *Dictyostelium*', *J. Cell Biol.*, vol. 74, p. 901

P. Drochmans, C. Freudenstein, J.-C. Wanson, L. Laurent, T.W. Keenan, J. Stadler, R. Leloup and W.W. Franke (1978) 'Structure and Biochemical Composition of Desmosomes and Tonofilaments Isolated from Calf Muzzle Epidermis', *J. Cell Biol.*, vol. 79, p. 427

J.E.B. Fox and D.R. Phillips (1981) 'Inhibition of Actin Polymerization in Blood Platelets by Cytochalasins', *Nature*, vol. 292, p. 650

W.W. Franke, K. Weber, M. Osborn, E. Schmid and C. Freudenstein (1978a) 'Antibody to Prekeratin. Decoration of Tonofilament-Like Arrays in Various Cells of Epithelial Character', *Exp. Cell Res.*, vol. 116, p. 429

—, E. Schmid, M. Osborn and K. Weber (1978b) 'Different Intermediate-Sized Filaments Distinguished by Immunofluorescence Microscopy', *Proc. Nat. Acad. Sci.*, vol. 75, p. 5034

—, E. Schmid, S. Winter, M. Osborn and K. Weber (1979a) 'Widespread Occurrence of Intermediate-Sized Filaments of the Vimentin Type in Cultured Cells from Diverse Vertebrates', *Exp. Cell Res.*, vol. 123, p. 25

—, E. Schmid, K. Weber and M. Osborn (1979b) 'HeLa Cells Contain Intermediate-Sized Filaments of the Prekeratin Type', *Exp. Cell Res.*, vol. 118, p. 95

D.L. Gard, P.B. Bell and E. Lazarides (1979) 'Coexistence of Desmin and the Fibroblastic Intermediate Filament Subunit in Muscle and Nonmuscle Cells: Identification and Comparative Peptide Analysis', *Proc. Nat. Acad. Sci.*, vol. 76, p. 3894

— and E. Lazarides (1980) 'The Synthesis and Distribution of Desmin and Vimentin During Myogenesis *in vitro*', *Cell*, vol. 19, p. 263

N. Geisler, U. Plessmann and K. Weber (1982) 'Related Amino-Acid Sequences in Neurofilaments and Non-Neuronal Intermediate Filaments', *Nature*, vol. 296, p. 448

D.S. Gilbert, B.J. Newby and B.H. Anderton (1975) 'Neurofilament Disguise, Destruction and Discipline', *Nature*, vol. 256, p. 586

R.D. Goldman, B. Chojnacki and M.-J. Yerna (1979) 'Ultrastructure of Microfilament Bundles in Baby Hamster Kidney (BHK-21) Cells', *J. Cell Biol.*, vol. 80, p. 759

W.E. Gordon (1978) 'Immunofluorescent and Ultrastructural Studies of "Sarcomeric" Units in Stress Fibres of Cultured Non-Muscle Cells', *Exp. Cell Res.*, vol. 117, p. 253

A.I. Gottlieb, M.H. Heggeness and J.F. Ash (1979) 'Mechanochemical Proteins, Cell Motility and Cell-Cell Contacts: the Localisation of Mechanochemical Proteins Inside Cultured Cells at the Edge of an *in vitro* "Wound" ', *J. Cell Physiol.*, vol. 100, p. 563

B.L. Granger and E. Lazarides (1980) 'Synemin: A New High Molecular Weight Protein Associated with Desmin and Vimentin Filaments', *Cell*, vol. 22, p. 727

—, E.A. Repasky and E. Lazarides (1982) 'Synemin and Vimentin are Components of Intermediate Filaments in Avian Erythrocytes', *J. Cell Biol.*, vol. 92, p. 299

U. Gröschel-Stewart (1980) 'Immunochemistry of Cytoplasmic Contractile Proteins', *Int. Rev. Cytol.*, vol. 65, p. 193

J.H. Hartwig and T.P. Stossel (1975) 'Isolation and Properties of Actin, Myosin and a New Actin-Binding Protein in Rabbit Alveolar Macrophages', *J. Biol. Chem.*, vol. 250, p. 5696

— and T.P. Stossel (1979) 'Cytochalasin B and the Structure of Actin Gels', *J. Mol. Biol.*, vol. 134, p. 539

S. Hatano and F. Oosawa (1966) 'Isolation and Characterization of Plasmodium Actin', *Biochim. Biophys. Acta*, vol. 127, p. 488

— and M. Tazawa (1968) 'Isolation, Purification and Characterization of Myosin B from Myxomycete Plasmodium', *Biochim. Biophys. Acta*, vol. 154, p. 507

I.M. Herman and T.D. Pollard (1981) 'Electron Microscopic Localization of Cytoplasmic Myosin with Ferritin-Labelled Antibodies', *J. Cell Biol.*, vol. 88, p. 346

J.E. Heuser and M.W. Kirschner (1980) 'Filament Organisation Revealed in Platinum Replicas of Freeze-Dried Cytoskeletons', *J. Cell Biol.*, vol. 86, p. 212

H. Hoffmann-Berling and H.H. Weber (1953) 'Vergleich der Molilität von Zellmodellen und Muskelmodellen', *Biochim. Biophys. Acta*, vol. 10, p. 629

G. Isenberg, P.C. Rathke, N. Hulsmann, W.W. Franke and K.E. Wohlfarth-Botterman (1976) 'Cytoplasmic Actomyosin Fibrils in Tissue Culture Cells. Direct Proof of Contractility by Visualisation of ATP-induced Contraction in Fibrils Isolated by Laser Microbeam Dissection', *Cell Tissue Res.*, vol. 166, p. 427

H. Ishikawa, R. Bischoff and H. Holtzer (1968) 'Mitosis and Intermediate-Sized Filaments in Developing Skeletal Muscle', *J. Cell Biol.*, vol. 38, p. 538

—, R. Bischoff and H. Holtzer (1969) 'Formation of Arrowhead Complexes with Heavy Meromyosin in a Variety of Cell Types', *J. Cell Biol.*, vol. 43, p. 312

R.E. Kane (1975) 'Preparation and Purification of Polymerized Actin from Sea Urchin Egg Extracts', *J. Cell Biol.*, vol. 66, p. 305

— (1976) 'Actin Polymerization and Interaction with Other Proteins in Temperature-Induced Gelation of Sea Urchin Egg Extracts', *J. Cell Biol.*, vol. 71, p. 704

M.W. Klymkowsky (1981) 'Intermediate Filaments in 3T3 Cells Collapse After Intracellular Injection of a Monoclonal Anti-Intermediate Filament Antibody', *Nature*, vol. 291, p. 249

T.E. Kreis, K.H. Winterhalter and W. Birchmeier (1979) '*In vivo* Distribution and Turnover of Fluorescently Labelled Actin Micro-injected into Human

Fibroblasts', *Proc. Nat. Acad. Sci.*, vol. 76, p. 3814
— and W. Birchmeier (1980) 'Stress Fibre Sarcomeres of Fibroblasts are Contractile', *Cell*, vol. 22, p. 555
E. Lazarides (1975) 'Tropomyosin Antibody: the Specific Localization of Tropomyosin in Non-Muscle Cells', *J. Cell Biol.*, vol. 65, p. 549
— (1976) 'Aspects of the Structural Organisation of Actin Filaments in Tissue Cultured Cells', in R. Goldman, T. Pollard and J.L. Rosenbaum (eds.), *Cell Motility* Book A (Cold Spring Harbor), pp. 347-60
— (1980) 'Intermediate Filaments as Mechanical Integrators of Cellular Space', *Nature*, vol. 283, p. 249
— (1981) 'Intermediate Filaments – Chemical Heterogeneity in Differentiation', *Cell*, vol. 23, p. 649
— and K. Burridge (1975) 'Alpha Actinin: Immunofluorescent Localization of a Muscle Structural Protein in Non-Muscle Cells', *Cell*, vol. 6, p. 289
W.H. Lewis and M.R. Lewis (1924) 'Behaviour of Cells in Tissue Cultures', in E.V. Cowdry (ed.), *General Cytology* (University of Chicago Press), pp. 385-447
R.K.H. Liem, S.-H. Yen, G.D. Salomon and M.L. Shelanski (1978) 'Intermediate Filaments in Nervous Tissue', *J. Cell Biol.*, vol. 79, p. 637
D.C. Lin, K.D. Tobin, M. Grumet and S. Lin (1980) 'Cytochalasins Inhibit Nuclei-Induced Actin Polymerization by Blocking Filament Elongation', *J. Cell Biol.*, vol. 84, p. 455
S. Lin, D.C. Lin and M.D. Flanagan (1978) 'Specificity of the Effects of Cytochalasin B on Transport and Motile Processes', *Proc. Nat. Acad. Sci.*, vol. 75, p. 329
A.G. Loewy (1952) 'An Actomyosin-Like Substance from the Plasmodium of a Myxomycete', *J. Cell Comp. Physiol.*, vol. 40, p. 127
S. Maclean-Fletcher and T.D. Pollard (1980) 'Mechanism of Action of Cytochalasin B on Actin', *Cell*, vol. 20, p. 329
A. Morris and J. Tannenbaum (1980) 'Cytochalasin D Does Not Produce Net Depolymerization of Actin Filaments in HEp-2 Cells', *Nature*, vol. 287, p. 637
M. Osborn, T. Born, H.-J. Koitsch and K. Weber (1978) 'Stereo Immunofluorescence Microscopy. I. Three-Dimensional Arrangement of Microfilaments, Microtubules and Tonofilaments', *Cell*, vol. 14, p. 477
T.D. Pollard (1976) 'Cytoskeletal Functions of Cytoplasmic Contractile Proteins', *J. Supramol. Struct.*, vol. 5, p. 317
J.M. Sanger and J.W. Sanger (1980) 'Banding and Polarity of Actin Filaments in Interphase and Cleaving Cells', *J. Cell Biol.*, vol. 86, p. 568
M. Schliwa (1981) 'Proteins Associated with Cytoplasmic Actin', *Cell*, vol. 25, p. 587
— (1982) 'Action of Cytochalasin D on Cytoskeletal Networks', *J. Cell Biol.*, vol. 92, p. 79
J.A. Schloss and R.D. Goldman (1980) 'Microfilaments and Tropomyosin of Cultured Mammalian Cells: Isolation and Characterization', *J. Cell Biol.*, vol. 87, p. 633
J.V. Small (1981) 'Organisation of Actin in the Leading Edge of Cultured Cells: Influence of Osmium Tetroxide and Dehydration on the Ultrastructure of Actin Meshworks', *J. Cell Biol.*, vol. 91, p. 695
— and J.E. Celis (1978) 'Direct Visualisation of the 10 nm (100Å) Filament Network in Whole and Enucleated Cultured Cells', *J. Cell Sci.*, vol. 31, p. 393
P.E. Steig, H.K. Kimelberg, J.E. Mazurkiewicz and G.A. Banker (1980) 'Distribution of Glial Fibrillary Acidic Protein and Fibronectin in Primary Astroglial Cultures from Rat Brain', *Brain Res.*, vol. 199, p. 493
P.M. Steinert, W.W. Idler and R.D. Goldman (1980) 'Intermediate Filaments of Baby Hamster Kidney (BHK-21) Cells and Bovine Epidermal Keratinocytes

Have Similar Ultrastructures and Subunit Domain Structures', *Proc. Nat. Acad. Sci.*, vol. 77, p. 4534

T.P. Stossel and J.H. Hartwig (1976) 'Interactions of Actin, Myosin and a New Actin-Binding Protein of Rabbit Pulmonary Macrophages. II. Role in Cytoplasmic Movement and Phagocytosis', *J. Cell Biol.*, vol. 68, p. 602

T.-T. Sun, C. Shih and H. Green (1979) 'Keratin Cytoskeletons in Epithelial Cells of Internal Organs', *Proc. Nat. Acad. Sci.*, vol. 76, p. 2813

S.W. Tanenbaum (ed.) (1978) *Cytochalasins: Biochemical and Cell Biological Aspects* (Elsevier-North Holland, Amsterdam), *North Holland Research Monographs*, vol. 46

C.M. Thompson and L. Wolpert (1963) 'The Isolation of Motile Cytoplasm from *Amoeba proteus*', *Exp. Cell Res.*, vol. 32, p. 156

K. Wang and S.J. Singer (1977) 'Interaction of Filamin with F-Actin in Solution', *Proc. Nat. Acad. Sci.*, vol. 74, p. 2021

K. Weber, P.C. Rathke, M. Osborn and W.W. Franke (1976) 'Distribution of Actin and Tubulin in Cells and in Glycerinated Cell Models After Treatment with Cytochalasin B (CB)', *Exp. Cell Res.*, vol. 102, p. 285

A. Weeds (1982) 'Actin-binding Proteins – Regulators of Cellular Architecture and Motility', *Nature*, vol. 296, p. 811

R.R. Weihing (1976) 'Cytochalasin B Inhibits Actin-Related Gelation of HeLa Extracts', *J. Cell Biol.*, vol. 71, p. 303

N.K. Wessells, B.S. Spooner, J.F. Ash, M.O. Bradley, M.A. Ludueña, E.L. Taylor, J.T. Wrenn and K.M. Yamada (1971) 'Microfilaments in Cellular and Developmental Processes', *Science*, vol. 171, p. 135

H. Wisniewski, M.L. Shelansky and R.D. Terry (1968) 'Effects of Mitotic Spindle Inhibitors on Neurotubules and Neurofilaments in Anterior Horn Cells', *J. Cell Biol.*, vol. 38, p. 224

K.E. Wohlfarth-Botterman (1964) 'Cell Structures and Their Significance for Ameboid Movement', *Int. Rev. Cytol.*, vol. 16, p. 61

J.J. Wolosewick and K.R. Porter (1976) 'Stereo High-Voltage Electron Microscopy of Whole Cells of the Human Diploid Line WI-38', *Am. J. Anat.*, vol. 147, p. 303

— and K.R. Porter (1979) 'Microtrabecular Lattice of the Cytoplasmic Ground Substance. Artifact or Reality', *J. Cell Biol.*, vol. 82, p. 114

S.-H. Yen and K.L. Fields (1981) 'Antibodies to Neurofilament, Glial Filament and Fibroblast Intermediate Filament Proteins Bind to Different Cell Types of the Nervous System', *J. Cell Biol.*, vol. 88, p. 115

5 MICROTUBULES

The presence of elongated tubular structures in cilia and flagella was recognised during the early days of electron microscopy (Fawcett and Porter, 1954). But although the term 'microtubule' was introduced by Slautterback (1963) as a result of the examination of cells fixed in osmium tetroxide, it was the introduction of glutaraldehyde as a fixative, together with improvements in embedding and staining techniques, which led to a proper appreciation of the widespread occurrence of microtubules, apart from those in cilia, flagella, basal bodies and centrioles, as a feature of the cytoplasm of virtually all types of eukaryotic cell. It has also become clear that microtubules are more stable in some situations than in others; those in cilia and flagella are the most stable, while those found in the mitotic spindle and more generally throughout the cytoplasm are much more easily disrupted by, for example, low temperatures, high hydrostatic pressures, high Ca^{2+} concentrations, and by the so-called 'spindle poisons'. It is the general cytoplasmic variety of microtubule with which we are chiefly concerned, but we shall also take account of information derived from work on microtubules found elsewhere in the cell.

The Structure and Cytoplasmic Distribution of Microtubules

Within the cytoplasm, microtubules may occur singly or in bundles, or they may be disposed in more extensive organised arrays. Irrespective of these different patterns, they display (for the most part) a common basic structure in electron micrographs. In transverse section they usually appear as rings *circa* 24 nm in diameter; the electron translucent central region is about 15 nm across, and the electron dense wall about 5 nm thick (Figure 5.1). These measurements may well be underestimates, since the preparation of tissues for electron microscopy may cause shrinkage (Amos, 1979).

In longitudinal section they appear as two electron dense parallel lines (Figure 5.2); their total length is difficult to estimate but immunofluorescence reveals that individual microtubules may extend uninterruptedly for at least 50 μm in the cytoplasm of cultured fibroblasts (Weber and Osborn, 1979), and in nerve fibres it is possible that they

Figure 5.1: Electron micrograph of microtubules (in transverse section) in the mitotic spindle of the pennate diatom *Diatoma vulgare* (From K. McDonald, J.D. Pickett-Heaps, R. McIntosh and D.H. Tippit, *J. Cell Biol.*, vol. 74 (1977), pp. 377-88.)

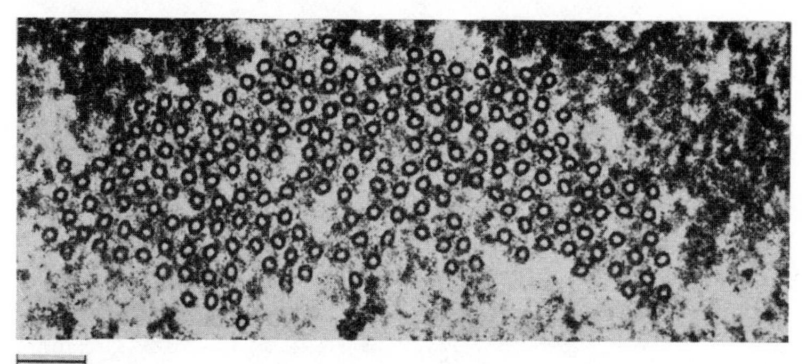

Figure 5.2: Electron micrograph (in longitudinal section) of microtubules isolated from erythrocytes of the crested newt *Triturus cristatus.* (From B. Bertolini and G. Monaco, *J. Ultrastruct. Res.*, vol. 54 (1976), pp. 59-67.)

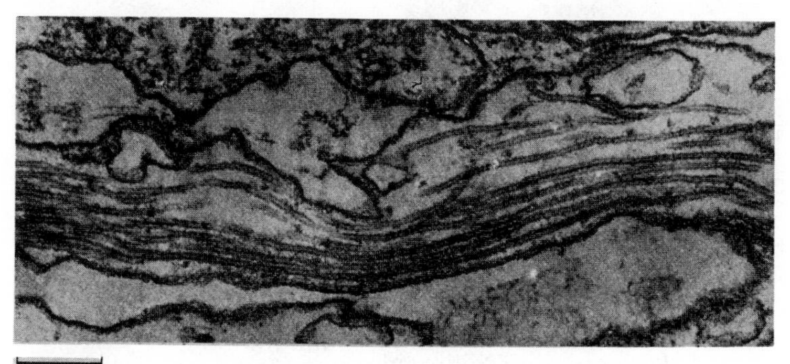

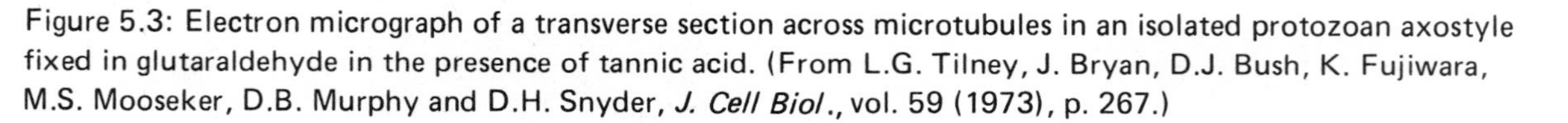

Figure 5.3: Electron micrograph of a transverse section across microtubules in an isolated protozoan axostyle fixed in glutaraldehyde in the presence of tannic acid. (From L.G. Tilney, J. Bryan, D.J. Bush, K. Fujiwara, M.S. Mooseker, D.B. Murphy and D.H. Snyder, *J. Cell Biol.*, vol. 59 (1973), p. 267.)

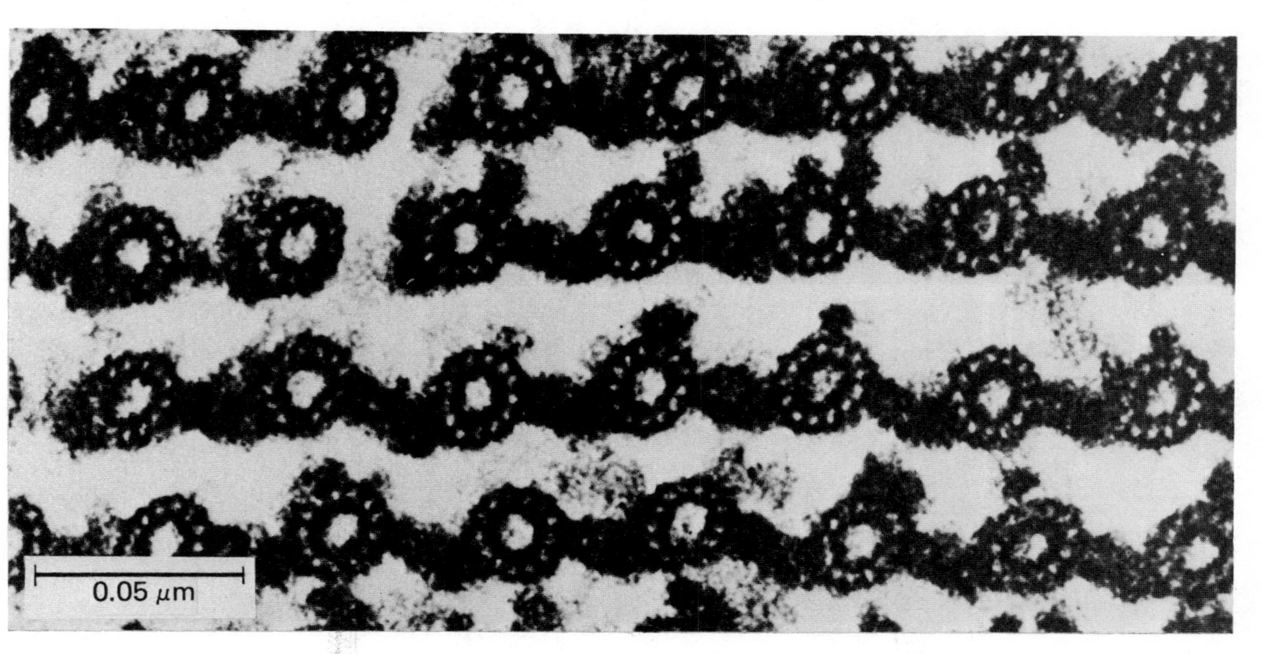

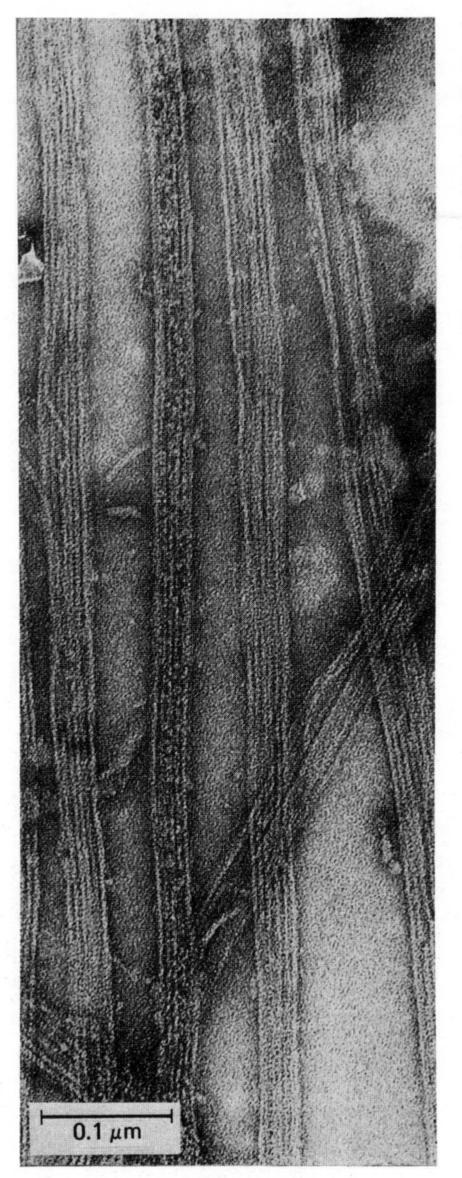

Figure 5.4: Electron micrograph of negatively stained microtubules isolated from erythrocytes of the crested newt *Triturus cristatus.* (From B. Bertolini and G. Monaco, *J. Ultrastruct. Res.*, vol. 54 (1976), pp. 59-67.)

may achieve lengths of several millimetres or even more (Porter, 1966). It was once thought that microtubules were straight, rigid structures, but it is clear from more recent immunofluorescence studies that they may undergo a considerable amount of bending.

Figure 5.5: Diagram of a microtubule showing how subunits of tubulin are arranged end-to-end to form longitudinal protofilaments. The subunits in adjacent protofilaments are staggered. (Redrawn from H. Stebbings and J.S. Hyams, *Cell Motility*, Longmans, 1979.)

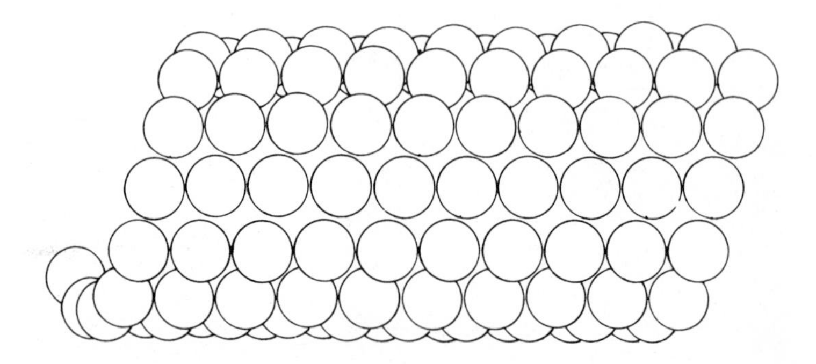

When tannic acid is included in the fixative, the wall as seen in transverse section shows particularly clearly the appearance of a 'ring of beads' (Figure 5.3); each 'bead' is 4 nm in diameter, and is in fact a cross-section of a protofilament running along the microtubule (Figure 5.4). In negatively stained specimens these protofilaments show an axial repeat of *circa* 8 nm (Figure 5.4) and in the majority of microtubules from a wide variety of sources there are usually 13 such protofilaments. Closer examination of individual protofilaments, using negative staining and optical diffraction techniques, reveals that each is constructed of a row of globular subunits about 4 nm in diameter (Figure 5.5). The 8 nm axial repeat is due to the end-to-end association of two such subunits. It is now clear that each subunit consists of a molecule of the protein, appropriately named 'tubulin', which is the major component of microtubules.

Although the electron microscope has provided useful information about the basic structure of microtubules, it is to the technique of immunofluorescence that we owe much of our knowledge of their distribution within the cytoplasm. The application of anti-tubulin antibodies to monolayers of cells *in vitro* has been particularly rewarding, since the cells tend to be well spread, with a minimal thickness of cytoplasm, making it much easier to appreciate the pattern of disposition of the microtubules throughout the cytoplasm. A wide variety of cultured cells has been studied in this way, using direct or indirect immunofluorescence (see Weber and Osborn, 1979, for a review). The method has a greater resolving power than might be expected; it is possible to detect single microtubules with the fluorescence microscope (Osborn *et al.*, 1978), but when microtubules lie closer to each other than 200-250 nm, they will not be resolved as separate entities because of the limit of resolution of the light microscope.

Figure 5.6: Distribution of microtubules in a cultured BHK fibroblast as revealed by indirect immunofluorescence with anti-tubulin antibody. (Courtesy of Dr Anne Woods.)

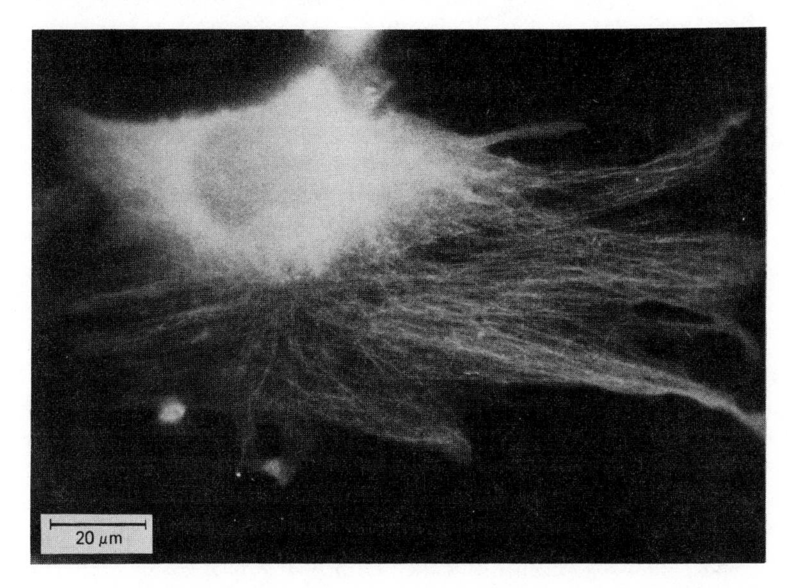

Immunofluorescence studies of a variety of different types of cell *in vitro* have shown that microtubules usually form extensive networks within the cytoplasm, often conforming approximately to the shape of the cell (Figure 5.6). Such networks can best be examined with stereo immunofluorescence photomicrography which allows some appreciation of their disposition in three dimensions (Weber and Osborn, 1979). Thus it appears that microtubules are present at all levels within the cell, both above and below the nucleus. Individual tubules may show considerable bending, and extend for up to 50 μm. Microtubules often seem to radiate from one or more regions within the cell, which have come to be known as 'microtubule organising centres' (see p. 94).

During mitosis, this general cytoplasmic network disappears and is replaced by the microtubules of the mitotic spindle.

The Biochemistry of Microtubules

Tubulin, the predominant protein in microtubules, exists in aqueous solution as a dimer with a sedimentation coefficient of 6*S*. When denatured, the dimer yields two monomers, each with a diameter of *circa* 4 nm, and it is these monomers which are visible as the globular subunits in negatively stained microtubules. Electrophoresis in SDS-polyacrylamide gels has shown that the two monomers migrate as separate bands called α and β tubulin, the latter having the greater mobility. Although these two tubulins have similar molecular weights (*circa* 50,000), they separate due to differences in their net charge. Analysis of their amino-acid sequences shows that α and β tubulin are closely related, and that the sequences from different species of animal are remarkably similar, suggesting that tubulin has been highly conserved in the course of evolution. For example incomplete sequence data show that the first 25 amino-acid residues are identical in α tubulin from chick embryo brain and from sea-urchin sperm (Ludueña and Woodward, 1973). Recently the complete amino-acid sequences of brain tubulins from pigs, chickens and rats have been established. These show that α tubulin has 450 or 451 amino-acid residues and β tubulin 445 residues, and confirm the similarities between the two types of tubulin and between tubulins derived from different animals (Krauhs *et al*., 1981; Ponstingl *et al*., 1981; Valenzuela *et al*., 1981). However, it is clear that many eukaryotes synthesise a number of variants of both α and β tubulins which differ slightly in their primary structure; thus there are at least four different α tubulins and two different β tubulins in pig brain (Krauhs *et al*., 1981; Ponstingl *et al*., 1981). It is not yet known whether these differences have any functional significance. The fact that, in most microtubules, α and β tubulins are present in a 1:1 ratio (e.g. Bryan and Wilson, 1971), suggests the possibility that each tubulin dimer is composed of an α and a β monomer, which is supported by experiments in which tubulin is treated with a reagent which forms intradimer cross-links, preventing separation of the two monomers. When chick brain tubulin is so treated, the resulting product consists predominantly of α-β dimers, indicating that, in solution, tubulin exists as the heterodimer rather than as homodimers (Ludueña, Shooter and Wilson, 1977).

These results, together with optical diffraction data, suggest that the protofilaments of microtubules consist of heterodimers of tubulin joined end-to-end so that α and β monomers alternate along the length of the filaments; the repetition of pairs of unlike monomers accounts for the *circa* 8 nm repeat visible along the protofilaments.

Drugs Which Act on Microtubules

There are several drugs which have been shown to prevent cells from completing mitosis by interacting with spindle microtubules or with tubulin; because they have this effect they are often referred to collectively as 'spindle poisons', but they have also been widely used to investigate other aspects of the function of microtubules unconnected with cell division. Colchicine, its derivatives, and the *Vinca* alkaloids, are probably the best known members of this group.

Colchicine

Obtained from the autumn crocus (*Colchicum autumnale*) and from other members of the Liliaceae, colchicine was used in the treatment of gout, perhaps as long ago as AD 550 (Hartung, 1954). It has been known to have an effect on cell division since the 1880s (Dustin, 1978), but it was not until it was possible to observe its action on dividing cells *in vitro* that it was established that colchicine halts mitosis at the premetaphase stage (e.g. Ludford, 1936). Later still it was shown to produce its effect by interfering with the formation and functioning of spindle microtubules, and it is now known to be capable of interfering with many other cellular activities involving microtubules, including the maintenance and alteration of cell shape, the extension of neurites, and some cytoplasmic movements. Further, it has been shown that it binds specifically to tubulin, each dimer having one binding site for colchicine. But the drug binds only to tubulin dimers in solution and not to those incorporated into the protofilaments of microtubules (see Ludueña, 1979, for a review).

A number of derivatives of colchicine have been used in investigations of microtubular functions (see Dustin, 1978, for a review); colcemid (desacetyl-N-methyl-colchicine) is perhaps the best known of these.

Vinblastine

This alkaloid can be extracted from the periwinkle (*Vinca rosea*) and, like colchicine, it interferes with microtubules in the mitotic spindle and elsewhere in the cytoplasm. Cells treated with vinblastine often develop birefringent crystals in their cytoplasm and large tubular structures (about 35 nm in diameter) known as macrotubules may appear. Structures resembling the protofilaments of microtubules are found in the crystals and in the macrotubules, and both types of structure are known to consist of complexes of tubulin and vinblastine. There are evidently two binding sites for vinblastine on each dimer of tubulin, both of which are different from the colchicine binding site (see Ludueña, 1979). Vincristine is one of several analogues of vinblastine which have also been used.

Other Spindle Poisons

Other antimitotic compounds which affect microtubules have been identified (Dustin, 1978; Ludueña, 1979). The following have found significant application in research on microtubules.

Podophyllotoxin is an alkaloid from the may apple (*Podophyllum peltatum*) which binds to tubulin; its binding site overlaps with, but is not identical to, the colchicine binding site. Yet another plant alkaloid, maytansine, apparently competes for the same sites on tubulin dimers as vinblastine. The antifungal drug griseofulvin has effects very similar to those produced by colchicine; its binding site has not yet been identified, but it is possible that it influences the link between 'microtubule associated proteins' (MAPs) and tubulin (see below). More recently a new class of antifungal drugs, the benzimidazole carbamates, have been shown to bind to tubulin (Davidse and Flach, 1977) and are increasingly being used to study microtubule functions, particularly in lower organisms.

The Assembly of Microtubules in the Living Cell

Having studied mitosis in a variety of animal and plant cells, Inoué proposed that the 'fibres' (microtubules) of the mitotic spindle exist in a dynamic equilibrium with subunits in the cytoplasm (reviewed by Inoué and Sato, 1967). Using polarised light microscopy he demonstrated that spindle fibres are labile structures that disappear if exposed to colchicine, to low temperatures, or to high hydrostatic pressure, reappearing when the colchicine, etc., is withdrawn. The fact that the spindle can reappear in the absence of protein synthesis shows that the microtubules must reform from existing cytoplasmic units (Inoué and Sato, 1967), and there is now no doubt that these microtubules exist in a state of equilibrium with soluble dimeric tubulin subunits in the cytoplasm. The same is true for other cytoplasmic microtubules; for instance the elongation of lens epithelial cells involves the assembly of microtubules, but the initial stages of this elongation can occur without synthesis of new protein and without any change in the total content of tubulin in the cells (see Piatigorsky, 1975, for a review). An increase in the number of cytoplasmic microtubules is also seen when Chinese hamster ovary (CHO) cells *in vitro* assume a flattened 'fibroblastic' morphology in response to treatment with cAMP, and again this is independent of protein synthesis (Patterson and Waldren, 1973). Similarly, the elongation of the neurite (axon) of a neuroblastoma cell in culture demands the formation of microtubules but can occur without synthesis of protein and does not cause any change in the total amount of tubulin in the cell (Morgan and Seeds, 1975). These, and other similar findings (see Fulton and Simpson, 1979), all indicate that there is a pool of tubulin dimers in the cytoplasm of many types of cell which, under appropriate conditions, can be reversibly assembled into microtubules. The actual proportion of the tubulin in a cell which is present as microtubules may be as little as 2 per cent or as much as 90 per cent, depending on the type of cell and its activity (see Fulton and Simpson, 1979, for a review). In a given cell, this proportion may be reduced by colchicine, low temperature, or increased hydrostatic pressure, all of which promote the disassembly of microtubules. Removal of these factors, or exposure of the cells to heavy water (D_2O), shifts the equilibrium between polymerised and unpolymerised tubulin back in favour of polymerisation to increase the numbers of microtubules.

Certain microtubules, the so-called stable microtubules of cilia and flagella (see above), are resistent to colchicine. From what we know of the action of colchicine it is clear that they cannot therefore be in a state of dynamic equilibrium with tubulin dimers in the cytoplasm. Tubulin extracted from these stable microtubules shows the same colchicine binding properties as tubulin from labile cytoplasmic microtubules. In addition, the assembly of microtubules involved in the regeneration of cilia and flagella is sensitive to colchicine. It appears, therefore, that there is nothing inherently different about the tubulin in these stable microtubules, but rather that their structure is stabilised in some way that prevents them from participating in the equilibrium between labile microtubules and cytoplasmic tubulin dimers.

Microtubule Organising Centres

Many cellular activities require not only that microtubules are assembled from the pool of cytoplasmic tubulin but also that this assembly should occur at the appropriate place within the cell; the production of the mitotic spindle is an obvious example. There is evidence that, in many cases, the formation of microtubules is associated with specific structures that act as nucleating sites for tubule assembly; although these structures take different forms, they are known collectively as microtubule organising centres (MTOCs) (Pickett-Heaps, 1969). Thus the basal bodies associated with cilia and flagella apparently act as MTOCs for the microtubules in these structures, and the centrosomes (centrioles plus centrosphere) act as MTOCs for spindle microtubules during mitosis.

Using anti-tubulin antibodies and immunofluorescence, Brinkley *et al.* (1975) were able to show that in a number of cell lines *in vitro* most of the microtubules converged towards one or two regions close to the nucleus, which they presumed to represent MTOCs. Their number and position led these observers to suggest that they were centrosomes. Osborn and Weber (1976) found a similar situation in 3T3 cells and noted that, after disruption by colcemid, the microtubules reformed around one or two MTOCs. In contrast, Spiegelman *et al.* (1979) found an average of 12 MTOCs in undifferentiated neuroblastoma cells recovering from exposure to colcemid. These cells, however, have been shown also to have multiple centrioles, which probably accounts for the larger than expected number of MTOCs which they possess (Sharp *et al.*, 1981; Brinkley *et al.*, 1981). Brinkley and his colleagues (1981) have confirmed that during interphase a number of different mammalian cells in culture contain a single MTOC; this is associated with the centrosome and is duplicated during the S phase of the cell cycle. They conclude that the centrosome constitutes the primary MTOC of interphase cells, and that this organelle acts as a template for the initiation and assembly of specific microtubular arrays.

The Assembly of Microtubules in Vitro

In 1972, Weisenberg established that, under suitable conditions, microtubules would form spontaneously *in vitro* when tubulin extracted from rat brain was warmed to 37°C, and modifications of his technique have been used extensively to study the mechanism of formation of microtubules. The prerequisites for the spontaneous polymerisation of tubulin include a minimum protein concentration, a pH of about 6.9, the presence of guanosine triphosphate and magnesium ions, and the absence of calcium ions. Polymerisation is inhibited by low temperatures, by colchicine, and by millimolar concentrations of calcium ions (reviewed by Raff, 1979). Characteristically, the formation of microtubules occurs after a short delay, which is assumed to be due to an initiating process in which nucleating sites are formed. In fact, in some tubulin extracts polymerisation will not start unless they are 'seeded' with fragments of microtubules which then act as nucleating sites. Extracts which are capable of spontaneous polymerisation have frequently been found to contain tubulin oligomers, often in the form of rings, double rings or spirals, and in some cases these have proved to be necessary to initiate microtubule formation (Borisy *et al.*, 1972).

The study of microtubule formation *in vitro* in brain extracts has revealed that other proteins in addition to tubulin may be necessary, and these have come to be known as the 'microtubule associated proteins' or MAPs. They are divisible into two groups. One group has a high molecular weight (>250,000 daltons) (Sloboda *et al.*, 1976), the other has a lower MW (55,000-70,000 daltons) and these are known as the *tau* proteins (Weingarten *et al.*, 1975). Both groups are found in close association with tubulin and remain in association with it through several cycles of polymerisation and depolymerisation. The number of microtubules which are formed, and their rate of assembly, is in some cases proportional to the amount of MAPs present, but the significance of this is obscure. One suggestion is that the high MW proteins may stabilise and preserve lateral interactions between tubulin dimers in adjacent protofilaments while the *tau* proteins may provide longitudinal stability in the filaments (Scheele and Borisy, 1979). However, the speculative nature of such suggestions is underlined by the following observations:

1. Purified tubulin can form microtubules in the absence of MAPs if seeded with microtubule fragments (Murphy *et al.*, 1977).
2. At high concentrations of magnesium ions, pure tubulin will form microtubules in the absence of MAPs (Herzog and Weber, 1977).
3. MAPs can be replaced by a number of non-specific compounds such as ribonuclease A, diethylaminoethyl-dextran, or poly-L-lysine, which will promote the formation of microtubules in preparations of pure tubulin (Scheele and Borisy, 1979).

Thus the weight of the evidence favours the possibility that, at least *in vitro*, MAPs are not essential for the formation of microtubules from brain tubulin.

The formation of microtubules in cellular extracts derived from sources other than brain has not yet been studied extensively. Bulinski and Borisy (1979) have reported the formation of microtubules in extracts of cultured HeLa cells and have identified two groups of proteins that co-purify with HeLa tubulin through several cycles of microtubule assembly and disassembly. These HeLa MAPs have molecular weights of approximately 210,000 and 120,000 daltons, respectively, and both stimulate the polymerisation of tubulin purified from HeLa cells and from brain. Immunofluorescence with antibodies raised against these proteins reveals that they are both present along the lengths of the microtubules of cultured HeLa cells (Bulinski and Borisy, 1980a). In addition, proteins antigenically similar to the 210,000 dalton HeLa MAP are associated with the microtubules of a wide variety of different cells from humans and other primates (Bulinski and Borisy, 1980b). Significantly perhaps, such proteins were not detected in cells from a number of non-primate sources, and neither of the antibodies prepared against HeLa MAPs cross-reacted with either of the high molecular weight or *tau* MAPs purified from brain (Bulinski and Borisy, 1980a, 1980b). This may imply that different groups of organisms, and perhaps even different tissues, have different MAPs. The significance of such differences has yet to be explained.

The fact that tubulin extracts can be prepared under conditions which prevent the assembly of microtubules unless nucleating sites are added provides a useful technique for testing the ability of presumptive MTOCs to promote the assembly of tubules *in vitro*. Thus the basal bodies of cilia and flagella, chromosomal kinetochores, and centrosomes isolated from cultured cells have all been shown to act as nucleating sites for *in vitro* microtubule production (see Raff, 1979, for a review). Similar techniques have allowed the elongation of microtubules to be studied *in vitro*. When flagellar microtubules (axonemes) are used to promote tubule formation in preparations of brain tubulin, the microtubules elongate primarily from what had been *in vivo* the distal end of the axoneme (Allen and Borisy, 1974). The preferred direction of elongation is thought to reflect the intrinsic polarity of microtubules resulting from the arrangement of the asymmetrical tubulin dimers along the protofilaments. It has become apparent that tubulin subunits can, in fact, be added or removed at both ends of a microtubule; the preferential growth in length at one end is the result of the greater rate constant for net tubulin addition at that end (the 'fast' or 'plus' end) as compared with the other ('slow' or 'minus') end, where the net tubulin loss is greater. Under steady state conditions it is proposed that there is a balance between net tubulin addition at the 'fast' end and net tubulin removal at the 'slow' end (see Margolis and Wilson, 1981, for a review). Such a situation would result in a movement (or treadmilling) of tubulin subunits along the length of the microtubule from the 'fast' to the 'slow' end. The rate of this movement has been variously estimated as 0.7 μm/hr to 50 μm/hr *in vitro*; it has not been demonstrated *in vivo* (Margolis and Wilson, 1981).

This intrinsic polarity of microtubules could be important in understanding their functions, but it is only recently that methods have been developed to demonstrate their polarity in intact cells. Heidemann and McIntosh (1980) have shown that under certain conditions microtubules formed *in vitro* with basal bodies or fragments of spindle microtubules as nucleating sites were 'decorated' with C-shaped arrays of laterally associated protofilaments. In cross-section these arrays were seen as hooks curving clockwise or anticlockwise around the microtubules. Further, they were able to establish that the direction of curvature of the hook depends on the polarity of the tubules; the hooks curve clockwise when viewed from the 'fast' end of the microtubule towards the 'slow' end. This can be expressed as a 'right-hand rule'; if the fingers of the right hand are curled in the direction of hook curvature, the thumb will point towards the 'slow' end of the microtubule (Euteneuer and McIntosh, 1980). It has recently become possible to determine the polarity of existing microtubules within cells, as well as of those made *in vitro*; for example in cultured PtK_2 and in HeLa cells the microtubules in one half of the mitotic spindle can all be shown to have the same polarity, and the two halves of the spindle are organised so that their 'fast' ends overlap at the equator (Euteneuer and McIntosh, 1980). It seems reasonable to predict that the ability to determine the polarity of microtubules will be important in helping us to understand how they function in the living cell.

References

C. Allen and G.G. Borisy (1974) 'Structural Polarity and Directional Growth of Microtubules of *Chlamydomonas* Flagella', *J. Molec. Biol.*, vol. 90, p. 381

L.A. Amos (1979) 'Structure of Microtubules', in K. Roberts and J.S. Hyams (eds.), *Microtubules* (Academic Press, London), pp. 1-64

G.G. Borisy, J.B. Olmsted and J.B. Klugman (1972) '*In Vitro* Aggregation of

Cytoplasmic Microtubule Subunits', *Proc. Nat. Acad. Sci.*, vol. 69, p. 2890
B.R. Brinkley, G.M. Fuller and D.P. Highfield (1975) 'Cytoplasmic Microtubules in Normal and Transformed Cells in Culture. Analysis by Tubulin Antibody Immunofluorescence', *Proc. Nat. Acad. Sci.*, vol. 72, p. 4981
—, S.M. Cox, D.A. Pepper, D.A. Wible, S.L. Brenner and R.L. Pardue (1981) 'Tubulin Assembly Sites and the Organisation of Cytoplasmic Microtubules in Cultured Mammalian Cells', *J. Cell Biol.*, vol. 90, p. 554
J. Bryan and L. Wilson (1971) 'Are Cytoplasmic Microtubules Heteropolymers?', *Proc. Nat. Acad. Sci.*, vol. 68, p. 1762
J.C. Bulinski and G.G. Borisy (1979) 'Self Assembly of Microtubules in Extracts of Cultured HeLa Cells and the Identification of HeLa Microtubule-Associated Proteins', *Proc. Nat. Acad. Sci.*, vol. 76, p. 293
— and G.G. Borisy (1980a) 'Immunofluorescence Localisation of HeLa Cell Microtubule-Associated Proteins on Microtubules *In Vitro* and *In Vivo*', *J. Cell Biol.*, vol. 87, p. 792
— and G.G. Borisy (1980b) 'Widespread Distribution of a 210,000 Molecular Weight Microtubule-Associated Protein in Cells and Tissues of Primates', *J. Cell Biol.*, vol. 87, p. 802
L.C. Davidse and W. Flach (1977) 'Differential Binding of Methyl benzimidazol-2-yl-carbamate to Fungal Tubulin as a Mechanism of Resistance to this Antimitotic Agent in Mutant Strains of *Aspergillus nidulans*', *J. Cell Biol.*, vol. 72, p. 174
P. Dustin (1978) *Microtubules* (Springer-Verlag, Berlin), p. 452
U. Euteneuer and J.R. McIntosh (1980) 'Polarity of Midbody and Phragmoplast Microtubules', *J. Cell Biol.*, vol. 87, p. 509
D.W. Fawcett and K.R. Porter (1954) 'A Study of the Fine Structure of Ciliated Epithelia', *J. Morphol.*, vol. 94, p. 221
C. Fulton and P.A. Simpson (1979) 'Tubulin Pools, Synthesis and Utilization', in K. Roberts and J.S. Hyams (eds.), *Microtubules* (Academic Press, London), pp. 117-74
E.F. Hartung (1954) 'History of the Use of Colchicine and Related Medicaments in Gout, with Suggestions for Further Research', *Ann. Rheum. Dis.*, vol. 13, p. 190
S.R. Heidemann and J.R. McIntosh (1980) 'Visualisation of the Structural Polarity of Microtubules', *Nature*, vol. 286, p. 517
W. Herzog and K. Weber (1977) '*In Vitro* Assembly of Pure Tubulin into Microtubules in the Absence of Associated Proteins and Glycerol', *Proc. Nat. Acad. Sci.*, vol. 74, p. 1860
S. Inoué and H. Sato (1967) 'Cell Motility by Labile Association of Molecules. The Nature of Mitotic Spindle Fibres and their Role in Chromosome Movement', *J. Gen. Physiol.*, vol. 50 (Suppl. *The Contractile Process*), p. 259
E. Krauhs, M. Little, T. Kempf, R. Hofer-Warbinek, W. Ade and H. Ponstingl (1981) 'Complete Amino Acid Sequence of β-Tubulin from Porcine Brain', *Proc. Nat. Acad. Sci.*, vol. 78, p. 4156
R.J. Ludford (1936) 'The Action of Toxic Substances Upon the Division of Normal and Malignant Cells *In Vitro*', *Arch. Exp. Zellforsch.*, vol. 18, p. 411
R.F. Ludueña (1979) 'Biochemistry of Tubulin', in K. Roberts and J.S. Hyams (eds.), *Microtubules* (Academic Press, London), pp. 65-116
— and D.O. Woodward (1973) 'Isolation and Partial Characterization of α and β Tubulin from Outer Doublets of Sea Urchin Sperm and Microtubules of Chick Embryo Brain', *Proc. Nat. Acad. Sci.*, vol. 70, p. 3594
—, E.M. Shooter and L. Wilson (1977) 'Structure of the Tubulin Dimer', *J. Biol. Chem.*, vol. 252, p. 7006
R.L. Margolis and L. Wilson (1981) 'Microtubule Treadmills – Possible Molecular

Machinery', *Nature*, vol. 293, p. 705
J.L. Morgan and N.W. Seeds (1975) 'Tubulin Constancy During Morphological Differentiation of Mouse Neuroblastoma Cells', *J. Cell Biol.*, vol. 67, p. 136
D.B. Murphy, K.A. Johnson and G.G. Borisy (1977) 'Role of Tubulin-Associated Proteins in Microtubule Nucleation and Elongation', *J. Molec. Biol.*, vol. 117, p. 33
M. Osborn and K. Weber (1976) 'Cytoplasmic Microtubules in Tissue Culture Cells Appear to Grow From an Organising Structure Towards the Plasma Membrane', *Proc. Nat. Acad. Sci.*, vol. 73, p. 867
—, R.E. Webster and K. Weber (1978) 'Individual Microtubules Viewed by Immunofluorescence and Electron Microscopy in the Same PtK_2 Cell', *J. Cell Biol.*, vol. 77, p. R27
D. Patterson and C.A. Waldren (1973) 'The Effect of Inhibitors of RNA and Protein Synthesis on Dibutyryl Cyclic AMP Mediated Morphological Transformations of Chinese Hamster Ovary Cells *In Vitro*', *Biophys. Biochem. Res. Comm.*, vol. 50, p. 566
J. Piatigorsky (1975) 'Lens Cell Elongation *In Vitro* and Microtubules', *Ann. N.Y. Acad. Sci.*, vol. 253, p. 333
J.D. Pickett-Heaps (1969) 'The Evolution of the Mitotic Apparatus: An Attempt at Comparative Cytology in Dividing Plant Cells', *Cytobios.*, vol. 1, p. 257
H. Ponstingl, E. Krauhs, M. Little and T. Kempf (1981) 'Complete Amino Acid Sequence of α-Tubulin from Porcine Brain', *Proc. Nat. Acad. Sci.*, vol. 78, p. 2757
K.R. Porter (1966) 'Cytoplasmic Microtubules and Their Functions', in G.E.W. Wolstenholme and M. O'Connor (eds.), *Principles of Biomolecular Organisation* (Ciba Foundation Symposium, J. and A. Churchill, London), pp. 308-45
E.C. Raff (1979) 'The Control of Microtubule Assembly *In Vitro*', *Int. Rev. Cytol.*, vol. 59, p. 1
R.B. Scheele and G.G. Borisy (1979) '*In Vitro* Assembly of Microtubules', in K. Roberts and J.S. Hyams (eds.), *Microtubules* (Academic Press, London), pp. 175-254
G.A. Sharp, M. Osborn and K. Weber (1981) 'Ultrastructure of Multiple Microtubule Initiation Sites in Mouse Neuroblastoma Cells', *J. Cell Sci.*, vol. 47, p. 1
D.B. Slautterback (1963) 'Cytoplasmic Microtubules: I. Hydra', *J. Cell Biol.*, vol. 18, p. 367
R.D. Sloboda, W.L. Dentler, R.A. Bloodgood, B.R. Telzer, S. Granett and J.L. Rosenbaum (1976) 'Microtubule-Associated Proteins (MAPs) and the Assembly of Microtubules', in R. Goldman, T. Pollard and J.L. Rosenbaum (eds.), *Cell Motility* Book C (Cold Spring Harbor), pp. 1171-1212
B.M. Spiegelman, M.A. Lopata and M.W. Kirschner (1979) 'Aggregation of Microtubule Initiation Sites Preceding Neurite Outgrowth in Mouse Neuroblastoma Cells', *Cell*, vol. 16, p. 253
P. Valenzuela, M. Quiroga, J. Zaldivar, W.J. Rutter, M.W. Kirschner and D.W. Cleveland (1981) 'Nucleotide and Corresponding Amino Acid Sequences Encoded by α and β Tubulin mRNAs', *Nature*, vol. 289, p. 650
K. Weber and M. Osborn (1979) 'Intracellular Display of Microtubule Structures Revealed by Indirect Immunofluorescence Microscopy', in K. Roberts and J.S. Hyams (eds.), *Microtubules* (Academic Press, London), pp. 279-313
M.D. Weingarten, A.H. Lockwood, S.-Y. Hwo and M.W. Kirschner (1975) 'A Protein Factor Essential for Microtubule Assembly', *Proc. Nat. Acad. Sci.*, vol. 72, p. 1858

R.C. Weisenberg (1972) 'Microtubule Formation *In Vitro* in Solutions Containing Low Calcium Concentrations', *Science*, vol. 177, p. 1104

6 CELL LOCOMOTION IN CULTURE

Fibroblasts have frequently been studied in investigations of cell spreading and locomotion, and most of what follows is, therefore, based on observations of this cell; occasional reference will, however, be made to work on other types of cell.

It should be emphasised that most studies have been concerned with the events that occur as cells, in a tissue culture medium, move over the two-dimensional surface of a coverslip or the bottom of a culture vessel. This is obviously very different from the three-dimensional environment experienced by cells migrating within the body, and may well cause cells *in vitro* to display behaviour which they would not produce *in vivo*. It is therefore necessary to exercise considerable caution when attempting to predict behaviour *in vivo* from observations made *in vitro*. In an effort to overcome this problem, some investigators have studied the movements of cells cultured within three-dimensional matrices of collagen (e.g. Schor, 1980; Grinnell, 1982). As yet, however, cells in such cultures have not been studied so intensively as those in more conventional culture systems.

Spreading and Polarisation

When isolated fibroblasts are suspended in tissue culture medium, they generally assume an approximately spherical shape, but the surface of such a cell is not smooth. It is covered with numerous hemispherical protrusions known as blebs, and slender projections 0.1-0.2 μm in diameter and 5-30 μm in length known as filopodia (or microspikes) (Figure 6.1a). If the cell makes contact with a flat substratum to which it can adhere, the surface in contact with the substratum at first flattens slightly and the cell then spreads gradually to cover an increasing area of the substratum. In the course of spreading, existing and newly-formed filopodia touch and adhere to the substratum. In addition, thin sheet-like protrusions called lamellae are extended from the cell surface and also stick to the substratum. These lamellae often spread to fill the gap between two or more adjacent filopodia already in contact with the substratum. At first, spreading occurs radially around the entire circumference of the cell so that, after 1-2 hours, the thicker central region is

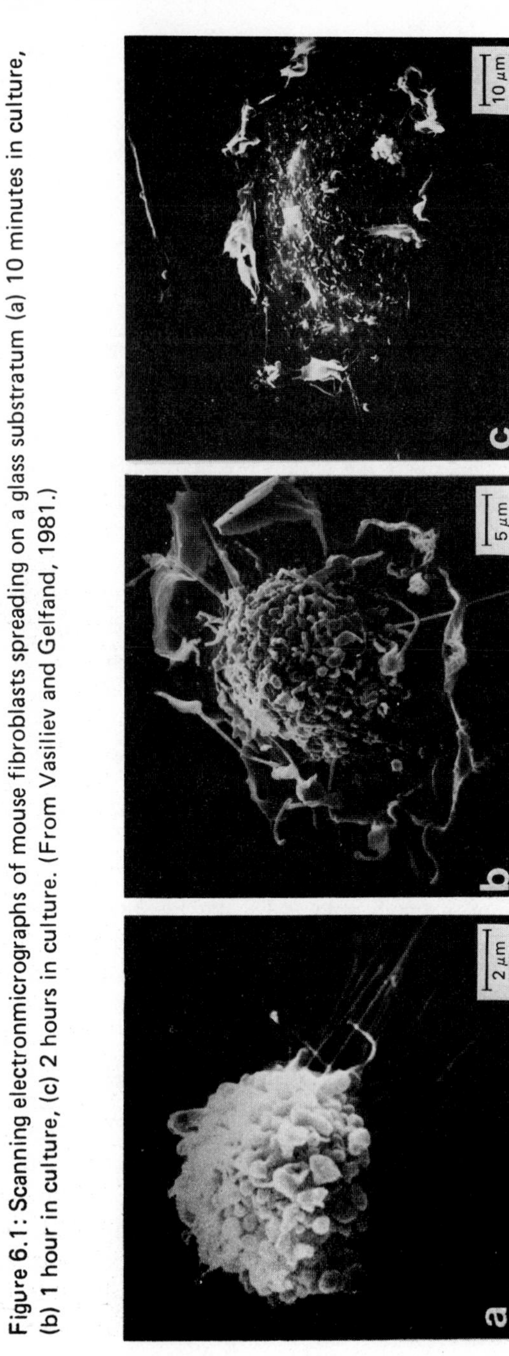

Figure 6.1: Scanning electronmicrographs of mouse fibroblasts spreading on a glass substratum (a) 10 minutes in culture, (b) 1 hour in culture, (c) 2 hours in culture. (From Vasiliev and Gelfand, 1981.)

surrounded on all sides by a zone of much thinner lamellar cytoplasm, giving the cell a characteristic 'fried-egg' appearance (Figure 6.1b) (see Vasiliev and Gelfand, 1981).

Figure 6.2: Phase contrast photomicrograph of a typical chick embryo heart fibroblast in culture, showing a leading lamella with ruffles

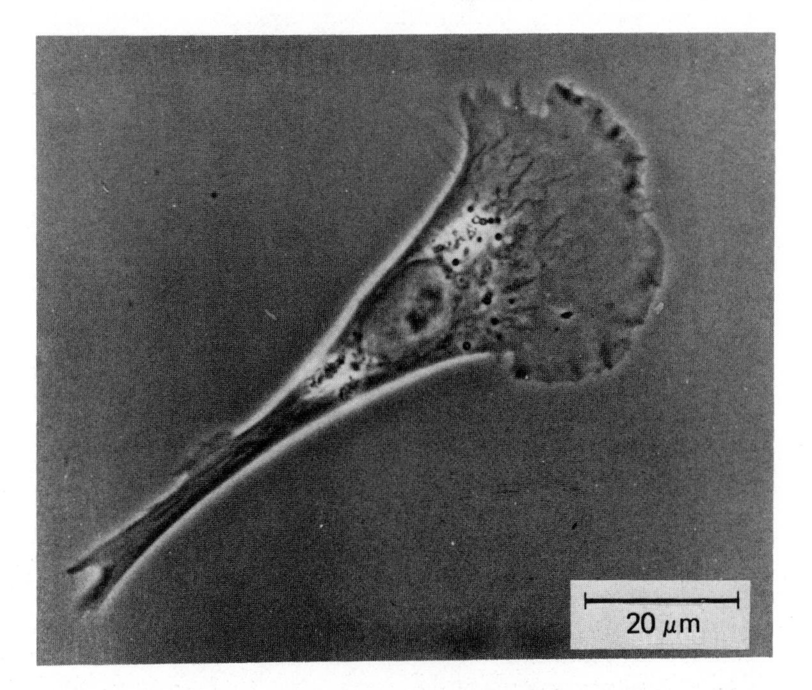

As spreading continues, the cell surface gradually becomes smoother as the number of blebs and filopodia decreases (Figure 6.1c); this suggests the possibility that the surface originally covering these structures becomes incorporated into the spreading lamellae (Erickson and Trinkaus, 1976). During the early stages of spreading, cells are essentially unpolarised; they do not move in any particular direction but spread in all directions simultaneously. After several hours in contact with a substratum, however, fibroblasts usually develop a polarised morphology; lamellar activity becomes restricted to a limited region of the cell margin and as a result the cells lose their roughly circular 'fried-egg' shape and often become triangular. A fibroblast cultured from embryonic chick heart is shown in Figure 6.2. Typically, these cells have a broad sheet of lamellar cytoplasm at their leading edge and their sides converge to a narrower 'tail' at the rear. Such cells are extensively flattened on the substratum; vertical sections show that the nuclear region, usually the thickest part of the cell, may be only 2-3 μm thick and the lamellar cytoplasm is usually 0.5 μm thick or less (Abercrombie *et al.*, 1971). The advancing margin of the cell frequently displays a characteristic activity known as 'ruffling'. When seen under phase contrast optics, this consists of the formation of short, dark, wavy lines (ruffles) which usually form close to the front edge of the cell and move backwards towards the nucleus as the cell advances. This region of the cell has often been referred to as the 'ruffled membrane'. However, since the use of the word 'membrane' in this context could lead to confusion with the plasma membrane of the electron microscopist, the term 'leading lamella' is now preferred (Ingram, 1969). In addition to ruffles, filopodia may also be seen in association with the leading lamella.

The presence of a leading lamella polarises the locomotion of a cell because it engenders movement in a particular direction. It often appears as though the cell is being pulled along by the ruffling of its leading lamella. In general, each fibroblast has a single dominant leading lamella, but it is not a fixed structure and any part of the cell margin has the capacity to develop into the leading lamella. Indeed, minor areas of ruffling activity are frequently (but usually only briefly) evident in regions of the cell margin away from the leading lamella. The activity of the dominant leading lamella may be inhibited either spontaneously or by contact with other cells (see Chapter 7). When this occurs, a different part of the cell margin (often one already displaying minor ruffles) becomes the dominant leading lamella and the direction of locomotion is correspondingly altered. There may be some kind of competitive relationship between lamellae, a dominant one tending to suppress the formation of others (Abercrombie, 1980).

Fibroblasts *in vitro* move extremely slowly, often averaging only 0.5-1.0 μm/min. (Abercrombie *et al.*, 1970a), but such average figures disguise the fact that their locomotion is irregular. Detailed analysis shows that at some times the cells may be almost stationary while at others they may be moving at speeds equivalent to about 2.0 μm/min. (Abercrombie *et al.*, 1970a). A number of factors, including the nature of the substratum and contact with other cells, can influence the direction of locomotion, but even in the absence of such factors fibroblasts do not move randomly. An analysis of the movement of mouse fibroblasts at 2.5 hour intervals (using time-lapse cinemicrography – see

p. 31) showed that the cells moved neither in a totally random nor in a totally uniform manner. Instead, their pattern of movement lay between these two extremes, their locomotion tending to persist in a particular direction over successive time intervals (Gail and Boone, 1970), indicating that once a cell has begun to move in a certain direction, it will continue to do so for some time. Since the direction of locomotion is determined by the position of the leading lamella, it is clear that once a leading lamella has formed from a particular region of the cell margin it will have a tendency to persist there.

Cytoskeletal Changes Accompanying Cell Spreading

As we have seen (p. 61), well spread fibroblasts often contain complex arrays of microfilament bundles. Such bundles are not seen in cells in suspension or in cells newly attached to a substratum (see Goldman *et al.*, 1976), but they appear as spreading proceeds. Electron microscopy of 3T3 and BHK 21 cells has revealed that microfilament bundles appear rapidly in regions of the cytoplasm adjacent to the substratum even at the stage when only a limited amount of spreading has occurred (Goldman *et al.*, 1976). But in some mouse fibroblasts mere contact with the substratum is apparently not sufficient to promote the formation of these bundles and they only appear in the cytoplasm of the newly formed filopodia and lamellae that are extended during spreading (Bragina *et al.*, 1976). The submembranous cortical cytoplasm of cells in suspension contains a random meshwork of microfilaments and it is in this zone that microfilament bundles often appear first (Goldman *et al.*, 1976). These bundles can form in the absence of protein synthesis (Goldman and Knipe, 1973), so the cortical meshwork probably provides a pool of precursors, and contact with a substratum in some way initiates the conversion of some of this meshwork into microfilament bundles (Goldman *et al.*, 1976). An additional source of precursors might be present in the form of monomers of cytoplasmic G-actin which, presumably in association with one or more of the actin-binding proteins (see p. 70), could polymerise to form microfilaments which are then incorporated into bundles.

Immunofluorescence has been used to follow the changes in the distribution of actin and related proteins which occur during cell spreading. Lazarides (1976a) has shown that, when rat fibroblasts are spreading, the formation of actin-containing microfilament bundles is preceded by the appearance of a complex three-dimensional polygonal network of fibres reminiscent of a geodesic dome (Figure 6.3). The vertices of this network contain actin and α-actinin while the fibres connecting these vertices contain actin and tropomyosin. Fibres that are attached to the vertices at only one of their ends are also visible and these terminate at the edges of the spreading lamellae. These fibres are indistinguishable from the actin-containing fibres seen in the microfilament bundles of fully spread cells and, unlike fibres which pass from one vertex to another, contain both tropomyosin and α-actinin as well as actin (Lazarides, 1976a). Similar networks have been demonstrated in spreading cells from a gerbil fibroma but here the vertices of the network and the interconnecting fibres contain both actin and α-actinin. Ultrastructural observations of these cells have shown that the vertices of the networks are apparently attached to the inner surface of the cells' plasma membrane (Gordon and Bushnell, 1979). In muscle cells α-actinin is localised in the Z-discs and is thought to participate in the attachment of actin filaments to the plasma membrane (see Lazarides and Burridge, 1975). The presence of this protein at the vertices of networks in fibroblasts and the association of these vertices with the plasma membrane implies that α-actinin could have a similar role in non-muscle cells. Lazarides (1976b) has proposed that the vertices of the networks act as nucleation sites and cytoplasmic organisation centres for newly forming actin filament bundles. This suggestion is supported by HMM binding (see p. 48) which has shown that in gerbil fibroma cells about 70 per cent of the actin filaments associated with a particular vertex have the same polarity – the 'arrowheads' point away from the vertex, i.e. the filaments had their origin at the vertex (Gordon and Bushnell, 1979). Further, Lazarides (1976b) has produced evidence that the association of α-actinin with the vertices of the network precedes by about two hours the association of tropomyosin with the interconnecting fibres; perhaps by binding to the vertices α-actinin restricts, and may, therefore, control, the binding of tropomyosin to the fibres (Lazarides, 1976b). It seems likely, therefore, that α-actinin is implicated in the organisation of microfilament bundles and in their attachment to the cell membrane during spreading. A similar conclusion has been reached on the basis of observations on the distribution of α-actinin in the adhesive plaques that develop between the cell and its substratum (see below).

Cell spreading clearly involves the conversion of meshwork microfilaments into microfilament bundles. In some cases the formation of bundles is preceded by the transitory appearance of polygonal networks, and it is likely that these networks are structural precursors of the microfilament bundles, but how this conversion is effected, and whether network formation invariably precedes bundle formation, is not yet known.

Figure 6.3: A geodesic dome of fibres in a spreading rat fibroblast, revealed by indirect immunofluorescence with anti-tropomyosin antibody. (Courtesy of Prof. E. Lazarides.)

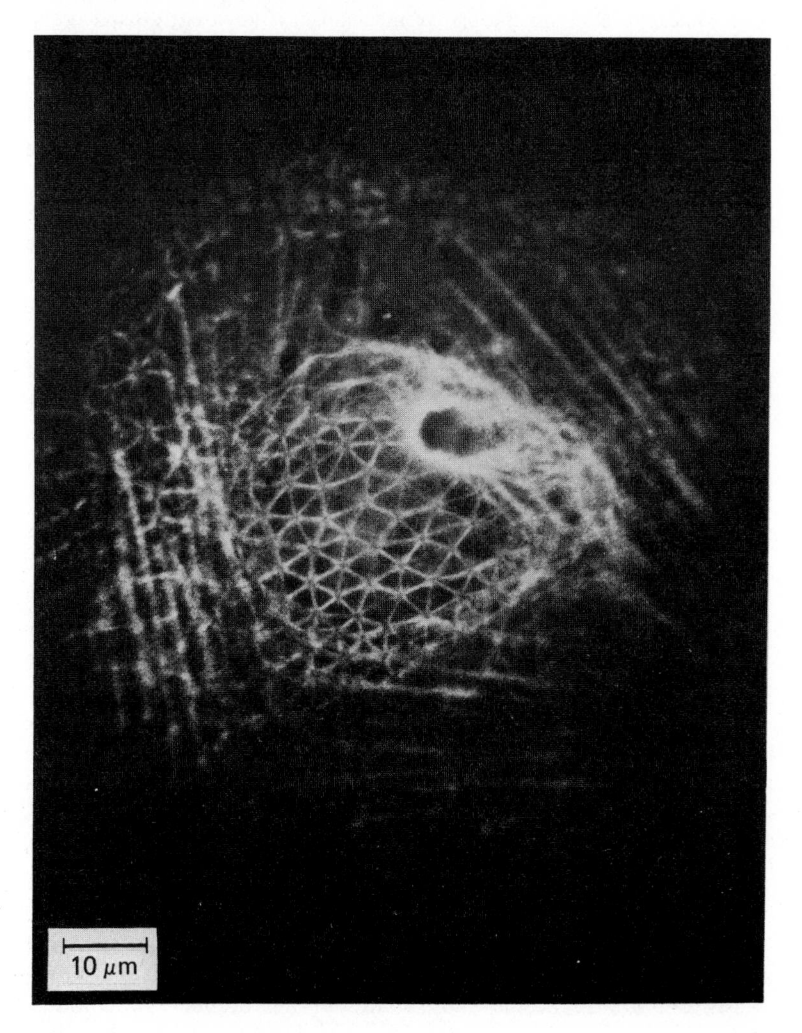

The possible contribution to the process of cell spreading made by these changes in the organisation of microfilaments remains unclear. In culture medium containing cytochalasin, mouse fibroblasts attach to the substratum but do not spread normally. Instead of forming lamellar protrusions, the cells develop a system of arborised processes, each 1-3 μm in diameter, lacking microfilament bundles (Bliokh *et al.*, 1980). Thus the cells can form protrusions in the presence of the drug, but some cytochalasin B-sensitive mechanism is apparently needed for the formation of normal lamellae.

Further changes in microfilament distribution accompany the subsequent polarisation of the cell; immunofluorescence has revealed that, before they become polarised, spreading cells contain both radially and tangentially disposed microfilament bundles, but as the cells polarise most bundles become orientated roughly parallel to the long axis of the cell. As a result, they lie approximately perpendicular to the leading lamella and parallel with the sides of the cell (see Vasiliev and Gelfand, 1981). Although mouse fibroblasts cannot spread normally in the presence of cytochalasin B, cells which have already spread can maintain a polarised morphology when subsequently treated with a concentration of the drug sufficient to disrupt their microfilament bundles (Vasiliev and Gelfand, 1981). Similarly, the microfilament bundles of polarised fibroblasts can be destroyed by treatment with various metabolic inhibitors without any consequent changes in cell shape (Bershadsky *et al.*, 1980). It seems, therefore, that the maintenance of a polarised morphology is not dependent on the continued presence of microfilament bundles.

Microtubules are also formed in the course of cell spreading and are required both to establish and to maintain polarisation. Immunofluorescence, using anti-tubulin antibodies, shows that in the earliest stages of spreading the lamellar cytoplasm is devoid of microtubules, but later an increasing number of microtubules become evident, extending radially from the nuclear region towards the margins of the spreading lamellae. As the margin advances with further spreading the microtubules elongate to keep pace with it. This elongation apparently occurs in a unidirectional manner towards the cell margin, possibly from one or more MTOCs (see p. 94) in the perinuclear region (Osborn and Weber, 1976; Vasiliev and Gelfand, 1981). In the presence of drugs which disrupt microtubules, such as colchicine or vinblastine, fibroblasts can attach to the substratum but microtubule formation is inhibited and arrays of cytoplasmic microtubules do not develop. The initial stages of spreading will occur under these conditions, but at a slower rate than is the case in medium lacking the drug (see Vasiliev and Gelfand, 1981), and the cells do not subsequently acquire a polarised morphology. Similarly, when established cultures of fibroblasts are treated with colcemid, their cytoplasmic microtubules disappear and the cells simultaneously lose their polarised shape. After addition of the drug to such cultures, lamellar activity and ruffling are no longer restricted to a limited portion of the cell margin but become widespread around the cell; the cells frequently revert to the 'fried-egg' morphology seen in the early stages of spreading, their locomotion is no longer polarised and, since they are now attempting to move in all directions simultaneously, they often come to a standstill. Careful analysis of time-lapse films has shown that, in the presence of colcemid, the directional persistence normally exhibited by moving fibroblasts is reduced, and the path taken by the cells, therefore, becomes more random (Gail and Boone, 1971).

It is clear from these observations that microtubules are important in inducing and maintaining cell polarisation and thus in directing cell locomotion, but we know little of the actual mechanisms involved. In polarised fibroblasts microtubules are frequently found close to relatively 'inactive' regions of the cell margin lacking lamellar activity, but they are comparatively rare in the more active region of the leading lamella. It has consequently been proposed that microtubules may in some way stabilise, and prevent the formation of lamellae in, regions of the cell margin with which they are associated. Hence they may restrict lamellar activity to a limited portion of the cell margin and so polarise the cell (see Vasiliev and Gelfand, 1981).

Activity of the Leading Lamella

The ruffling activity of the leading lamella is probably the most conspicuous feature of a moving fibroblast, and it has been subjected to detailed analysis. When cells are examined in profile (i.e. parallel with the substratum and at right angles to the direction of their movement), it is evident that ruffles usually form due to upfolding of the advancing edge of the lamella (Ingram, 1969). Part of this edge first protrudes forwards, usually parallel with the substratum, and then often rotates upwards at 2-3° per second until it is vertical (Harris, 1973a). These upturned portions of the leading lamella are visible as dark lines (ruffles) when the cell is viewed in the normal way with phase contrast (i.e. from above or below). A single ruffle usually involves only 5-10 μm of the edge of the lamella and several independent ruffles may coexist on the same lamella. Once formed, they usually move backwards towards the nucleus at about 2 μm per minute; this movement results both from the continued backward rotation of the ruffle about its base and from the gradual retreat of the base itself from the edge of the lamella (Harris, 1973a). Ruffles usually continue to rotate backwards beyond the vertical and eventually disappear as they merge with the upper surface of the lamella (Harris, 1973a; Abercrombie *et al.*, 1970b). On average, ruffles move backwards for about 1 μm, and are visible with phase contrast for less than a minute before they vanish (Abercrombie *et al.*, 1970b). Occasionally, however, more durable ruffles may move backwards for 20-30 μm before disappearing (Harris, 1973a). Although most ruffles are initiated at the edge of the leading lamella, they may also arise some distance behind the edge, when they are presumably formed by the more-or-less vertical protrusion of the upper surface of the lamella (Abercrombie *et al.*, 1970b). Abercrombie and his colleagues (1970a) have made a thorough study of the behaviour of the leading lamella in moving chick and mouse fibroblasts, and have found that, as a fibroblast advances, the edge of the leading lamella, relative to a fixed point on the substratum, is sometimes stationary, sometimes protruding forwards, and at other times is withdrawing. Adjacent regions of the edge, separated by as little as 6 μm, behave independently, so that one region may be protruding while a neighbouring one may be stationary or may be withdrawing. It seems that the edge of the leading lamella undergoes irregular cycles of protrusion and withdrawal, and the independent behaviour of contiguous parts of the edge indicates that these movements are generated relatively locally. The mean speeds of protrusion and withdrawal are both approximately the same (about 5 μm per minute); the cell nevertheless moves forward because a given region of the edge protrudes for about 30 per cent of the time but withdraws for only 20 per cent and is stationary for the remainder. In general, cells with highly active leading lamellae (in terms of the frequency and speed of their cycles of protrusion and withdrawal) move faster than those with less active lamellae (Abercrombie *et al.*, 1970a).

Ruffling is closely associated with this cyclical activity at the edge of the lamella. Although a minority (30 per cent) of cycles of protrusion and withdrawal are completed without ruffling, most do involve the formation of a ruffle. Significantly, most ruffles are associated with the withdrawal phase of the cycle, i.e. they appear in regions where the leading lamella is retreating rather than advancing (Abercrombie *et al.*, 1970b). This suggests that ruffling *per se* cannot be responsible for the forward locomotion of the cell, and it has been demonstrated that fibroblasts can move perfectly well when ruffling is inhibited by covering the cells with a viscous medium (Heaysman and Pegrum, 1982). The subjective impression, engendered by viewing time-lapse films, that

Figure 6.4: Electron micrograph of a complete chick embryo fibroblast sectioned longitudinally at right angles to the substratum. The cell was moving towards the right; the leading lamella is to the right of the nucleus and shows three lamellipodia (arrows). (From Abercrombie, Heaysman and Pegrum, 1971.)

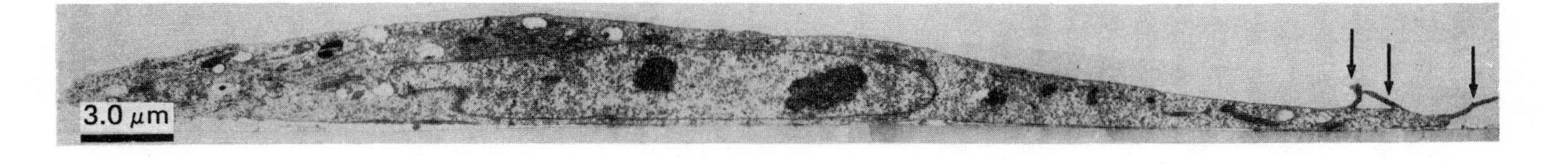

fibroblasts are somehow 'towed' along by the ruffling of their leading lamellae is misleading; the fact that ruffles usually occur when a previously protruded area of the leading lamella is withdrawn suggests that they represent an inefficient component of the locomotory process.

Figure 6.5: Electron micrograph of a vertical section of a lamellipodium arising from the leading lamella of a chick embryo heart fibroblast. (From Abercrombie, Heaysman and Pegrum, 1971.)

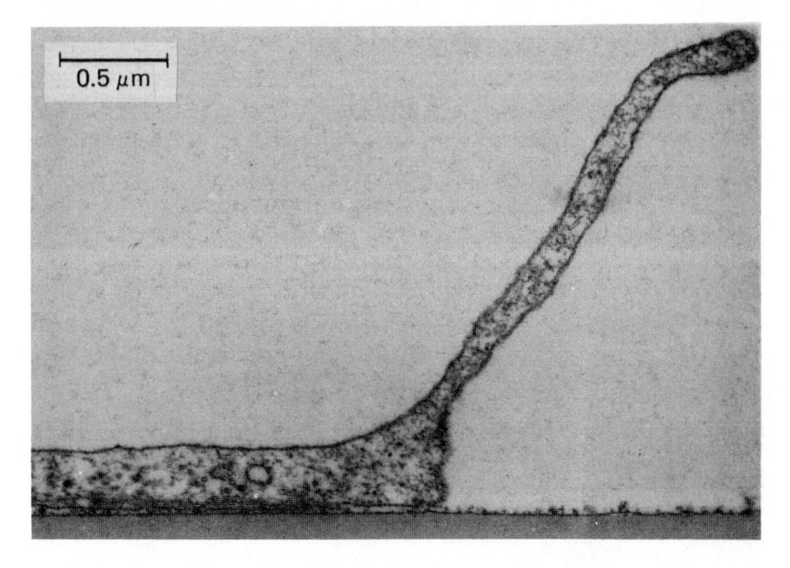

Electron microscopy has revealed that the protrusive phase of the cycle involves the formation of thin sheet-like projections called lamellipodia (Figure 6.4) (singular lamellipodium) from the edge of the leading lamella (Abercrombie *et al*., 1970b). Lamellipodia are usually *circa* 1.0-2.5 μm in length and have a rather uniform thickness of 110-160 nm (Figure 6.5); they are thus morphologically distinct from the leading lamellae from which they arise, which are rarely less than 200 nm thick (Abercrombie *et al*., 1971). In the protrusive phase of the cycle a lamellipodium pushes forwards parallel with, but usually not touching, the substratum. It may then establish an adhesion with the substratum, so causing the edge of the cell to advance, or it may withdraw. In the latter case, it may simply be reabsorbed into the leading lamella or, more often, it may fold upwards away from the substratum, forming a ruffle which then moves backwards over the surface of the leading lamella (Abercrombie *et al*., 1970b).

Leading lamellae are also produced at the edges of sheets of epithelial cells moving in culture, and Di Pasquale (1975a) has established that these epithelial lamellae show cycles of protrusion and withdrawal, and form ruffles, in much same way as those of fibroblasts. In addition, similar protrusive activity and ruffling are shown by the 'growth cones' that develop at the tip of nerve fibres extending in culture (see, for example, Wessels, 1982). Thus it seems likely that basically similar mechanisms underlie the behaviour of the leading lamellae of fibroblasts and epithelial cells, and of the growth cones of nerve fibres.

Adhesion to the Substratum

In order to move, cells must adhere to a substratum against which they can exert a rearward force. A number of different techniques have shown that the adhesions between a cell and its substratum are not uniformly distributed over its lower surface and involve only a small proportion of its area.

Regions of adhesion can be detected by the resistance that they offer to a fine probe introduced between a cell and its substratum, and this technique has revealed that, in moving fibroblasts, adhesions are confined to a rather narrow zone adjacent to the cell margin, and are almost never found beneath more central parts of the cell. They are especially abundant in those marginal regions which display protrusive activity, and infrequent in less active marginal areas (Harris, 1973b). Other techniques have been used to extend these observations and have provided confirmation that, in moving cells, the majority of cell-substratum adhesions are located beneath the leading lamella. Using the TEM, Abercrombie *et al.* (1971) found that most of the lower surface of leading lamellae of chick heart fibroblasts was separated from the substratum by a gap of about 60 nm. But in discrete patches, accounting for about 25 per cent of its area, this gap was reduced to *circa* 30 nm. The patches where the cell approached the substratum more closely had an average length and breadth of about a micrometre and the cytoplasm near the cell membrane in these regions contained filamentous material forming electron-dense 'plaques', often associated with bundles of microfilaments running obliquely backwards and upwards towards the nucleus (Figure 6.6). Abercrombie *et al.* (1971) considered that these regions represented areas of cell-substratum adhesion.

Considerable use has recently been made of interference reflection

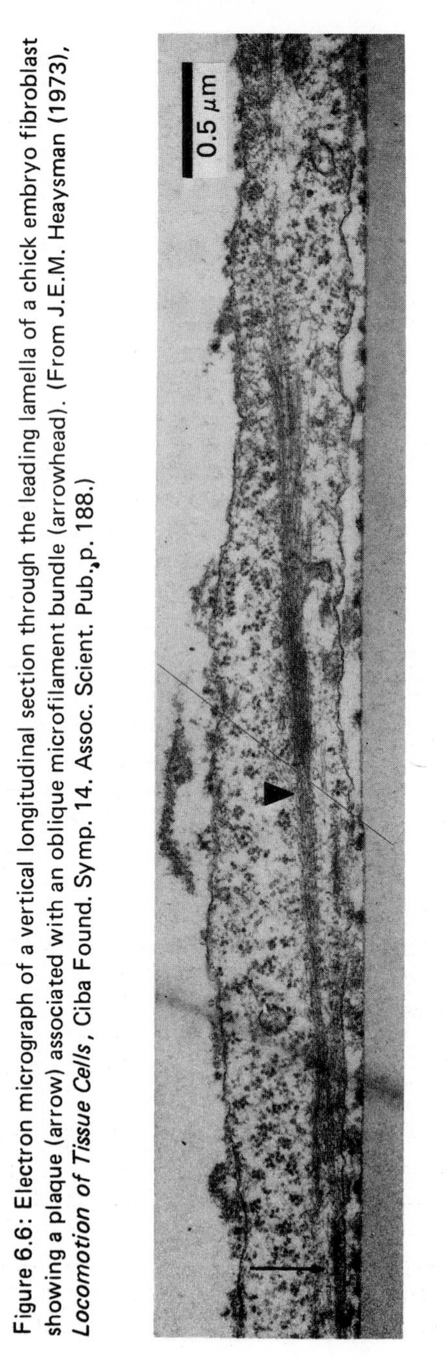

Figure 6.6: Electron micrograph of a vertical longitudinal section through the leading lamella of a chick embryo fibroblast showing a plaque (arrow) associated with an oblique microfilament bundle (arrowhead). (From J.E.M. Heaysman (1973), *Locomotion of Tissue Cells*, Ciba Found. Symp. 14. Assoc. Scient. Pub., p. 188.)

microscopy (see Chapter 3, p. 30) in the study of these adhesions. The images generated by this optical system are to a great extent determined by the size of the gap separating a cell from the substratum over which it is moving; although it allows sites of adhesion to be identified in living cells, and their behaviour during cell locomotion to be monitored, some caution is required when interpreting these images (see Chapter 3, p. 31). In white light, separations of three distinct types can be distinguished by differences in the intensity of the light reflected from the specimen. Areas in which the lower surface of the cell is very close (10-15 nm) to the substratum appear very dark or black and are known as 'focal contacts'; where the separation is greater (about 30 nm) the image appears grey and such regions are known as 'close contacts'. Where the separation is increased to 100 nm or more the image appears bright (Izzard and Lochner, 1976). Interference reflection images of a moving fibroblast (Figure 6.7) show a broad region of close contact lying beneath much of the leading lamella; most of the sites of focal contact also lie within this region but other focal contacts are often seen towards the trailing 'tail' of the cell. The focal contacts usually look like dark streaks 0.25-1.0 μm wide and 2.0-10.0 μm long, generally orientated parallel to the direction of movement of the leading lamella (Izzard and Lochner, 1976). Beneath the central part of the cell, near the nucleus, the image is bright and this area is not thought to be adherent to the substratum. Several lines of evidence suggest that both focal and close contacts represent areas of adhesion. As a fibroblast advances, focal and close contacts are formed close to its leading edge and both are apparently implicated in the adhesion of the leading lamella to the substratum (see below).

Close contacts alone can provide enough adhesion to permit locomotion, since some moving cells form only this type of contact, and seem to lack focal contacts (Couchman and Rees, 1979). When fibroblasts 'round up', either during mitosis or in response to low temperatures, their peripheral processes are withdrawn from the substratum towards the central rounded body of the cell, often leaving behind thin retraction fibres attaching the cell body to the substratum. When this rounding-up is viewed in the interference reflection microscope the distal ends of these fibres are found to be associated with either focal or, more rarely, close contacts, again suggesting that these contacts have adhesive properties (Abercrombie and Dunn, 1975; Izzard and Lochner, 1980). Similarly, if shearing forces are used to detach fibroblasts from their substratum, small fragments of their lower surfaces remain attached to the substratum and interference reflection micrographs show that the

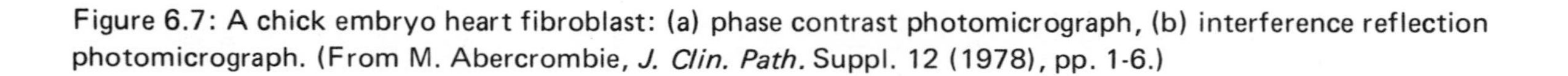

Figure 6.7: A chick embryo heart fibroblast: (a) phase contrast photomicrograph, (b) interference reflection photomicrograph. (From M. Abercrombie, *J. Clin. Path.* Suppl. 12 (1978), pp. 1-6.)

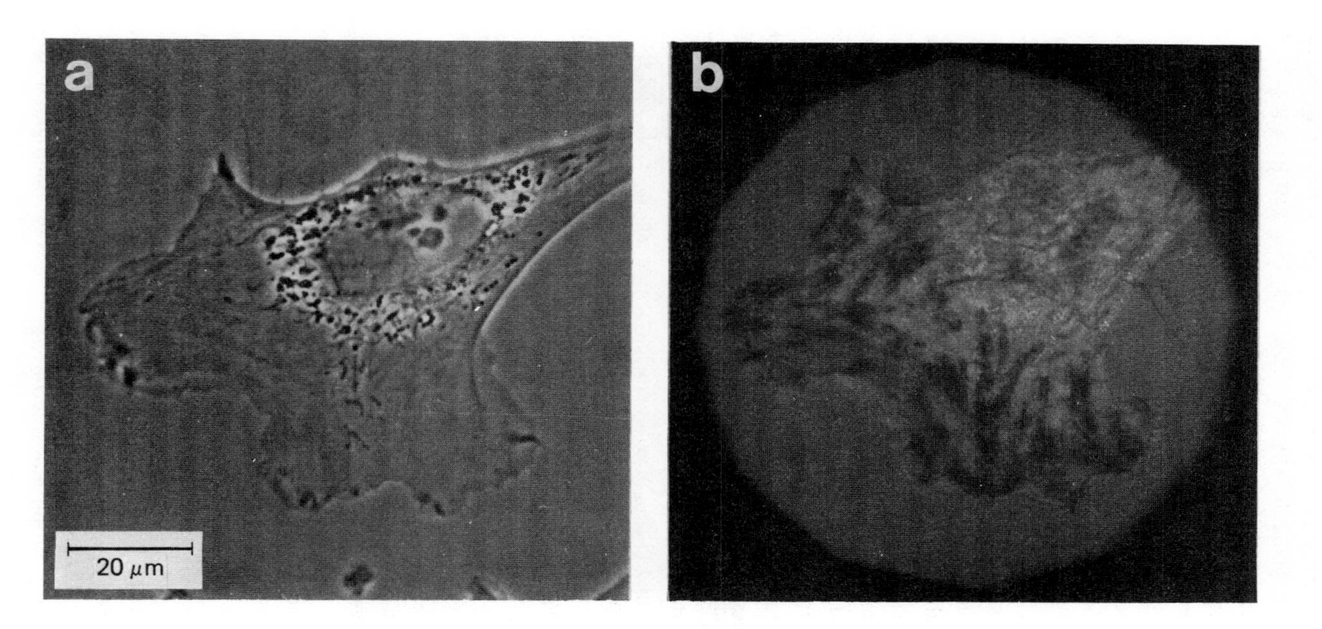

fragments occur at the sites of pre-existing focal contacts (Badley *et al.*, 1978). There is little doubt that both focal and close contacts are sites of cell-substratum adhesion, but what is their relationship to the adhesive 'plaques' described above? Early observations on moving fibroblasts by interference reflection microscopy revealed that focal contacts often occurred beneath prominent microfilament bundles visible by phase or Nomarski interference contrast (Abercrombie and Dunn, 1975; Izzard and Lochner, 1976). More detailed examination of fibroblasts using interference reflection microscopy and high voltage electron microscopy confirmed that focal contacts were invariably positioned at the ends of microfilament bundles (Heath and Dunn, 1978). Stereo viewing of pairs of HV electron micrographs showed that these bundles extended posteriorly and often obliquely upwards from the focal contacts to terminate in the region of the nucleus (Heath and Dunn, 1978). This arrangement is reminiscent of the association between microfilament bundles and adhesive plaques seen in transmission electron micrographs (see above) and it is now thought that adhesive plaques and focal contacts are one and the same structure (Abercrombie *et al.*, 1977).

Close contacts are not associated with well defined microfilament bundles, but the cytoplasm near the cell membrane in these regions contains a loose meshwork of microfilaments (Heath and Dunn, 1978).

The leading lamellae of the marginal cells at the edge of an advancing sheet of epithelium have a distribution of focal and close contacts similar to that seen in fibroblasts; much of the leading lamella in these cells forms a uniform close contact with the substratum and numerous focal contacts occur within this area. The non-marginal cells, however, have an interference reflection image that is markedly different; focal contacts are usually absent in these cells, which have only small areas of close contact, often interspersed with regions of greater separation between the cell and the substratum (Heath, 1982). This suggests that the marginal cells are mainly responsible both for the adhesion of epithelial sheets to the substratum and for exerting traction against the substratum during locomotion.

A number of techniques have been employed to characterise the components associated with focal contacts. There can be little doubt that the microfilament bundles which terminate in focal contacts are identical with those previously described (see Chapter 4, p. 61) and known to contain actin, myosin, α-actinin, tropomyosin and filamin. Comparisons of the images of identical cells obtained by interference reflection microscopy and stereo-immunofluorescence photomicrographs have confirmed that microfilament bundles terminating in focal contacts contain both actin and α-actinin (Wehland *et al.*, 1979). In addition, immunofluorescence has revealed that actin, myosin, α-actinin and tropomyosin are all present on the cytoplasmic side of focal contacts which remain attached to the substratum when the rest of the cell is detached by a stream of buffer (Badley *et al.*, 1978). A high concentration of α-actinin has been found in the region where a microfilament bundle terminates at a focal contact (Wehland *et al.*, 1979), which is evidence in favour of the proposal that this protein may act as a membrane anchorage protein for actin filaments in non-muscle cells (Lazarides and Burridge, 1975) (see Chapter 4, p. 60). However, the fact that α-actinin can be selectively removed from purified plasma membrane fractions without any significant loss of the actin present in these preparations suggests that α-actinin does not form a direct link between actin and the membrane (Burridge and McCullough, 1980). Another protein, vinculin, has also been found at focal contacts; this has a molecular weight of 130,000 daltons and was first isolated from smooth muscle. Immunofluorescence indicates that vinculin is even more sharply localised than α-actinin in the region of focal contacts (Geiger, 1979). When vinculin, purified and labelled with a fluorescent dye, is microinjected into fibroblasts, it accumulates selectively at the ends of microfilament bundles associated with adhesive plaques (Burridge and Feramisco, 1980), and it has been suggested that it may have a more direct role than α-actinin in associating actin filaments with the cell membrane (Geiger, 1979). *In vitro* both α-actinin and vinculin bind to purified F-actin, but with different results; α-actinin cross-links actin filaments to produce a gel in which it spaces out the individual actin filaments; vinculin on the other hand interacts with actin to form bundles of filaments with a paracrystalline substructure (Jockusch and Isenberg, 1981). Thus vinculin may promote the aggregation of individual actin filaments into microfilament bundles at focal contacts, whereas α-actinin may serve to 'fan out' the bundles a short distance from the focal contact, and may also stabilise them (Jockusch and Isenberg, 1981).

Fibronectin, a glycoprotein which is present in serum, and which is also secreted (in a somewhat different form) by fibroblasts in culture (see Yamada and Olden, 1978, for a review) has also been identified in association with the external face of the plasma membrane in the region of focal contacts. In some cases this protein has been found between the cell and the substratum at the sites of focal contacts (Hynes and Destree, 1978) but in other instances it has been reported to be absent from the immediate region of the contacts but has been found instead in association with the membrane adjacent to them (Chen and Singer, 1980). Close contacts, as previously mentioned, are usually associated with a cytoplasmic meshwork of microfilaments, but other than this we have little information about their nature.

Cell-substratum Contacts During Cell Locomotion

The interference reflection microscope has made it possible to watch the formation and fate of cell-substratum contacts in moving cells. Close and focal contacts are both formed near the leading edge of a fibroblast as it moves. The advance of the leading margin is paralleled by the advance of areas of close contact between the leading lamella of the cell and the substratum over which it is moving. Detailed analysis shows that the close contact advances only where a lamellipodium has first been protruded free of the substratum. The new area of close contact forms as the underside of this lamellipodium is lowered towards the substratum (Izzard and Lochner, 1980). The protrusion of a lamellipodium is evidently an essential preliminary for the subsequent formation of a close contact (Izzard and Lochner, 1980).

New focal contacts develop intermittently ahead of existing ones as the leading lamella advances, and once formed they remain stationary as the cell proceeds forward (Lochner and Izzard, 1973; Abercrombie and Dunn, 1975). Consequently, they come to be further and further behind the advancing cell margin, until eventually they, and their associated microfilament bundles, disappear. The life span of a focal contact is inversely related to the speed of movement of the cell and averages about 10 minutes (Lochner and Izzard, 1973). New focal contacts are formed only by an advancing leading lamella, mostly within 1.0-2.0 μm of the front edge of an existing close contact (Izzard and Lochner, 1980), although some do form independently of close contacts, in which case they develop as the result of either a filopodium or a lamellipodium bending towards the substratum (Izzard and Lochner, 1980). The combined use of Nomarski differential interference contrast microscopy and interference reflection microscopy has shown that in 90 per cent of cases the formation of focal contacts by chick embryo fibroblasts is preceded by the appearance in the cytoplasm of a short linear structure. With Nomarski optics this resembles a short bundle of microfilaments, and it appears over the site of the future focal contact some 10-70 seconds before the contact itself is detectable by interference reflection microscopy. Once the focal contact has been established, the linear feature elongates centripetally and after about 2 minutes a microfilament bundle running from the focal contact to the perinuclear region is apparent (Izzard and Lochner, 1980). These observations suggest strongly that the formation of a cytoplasmic specialisation, probably composed of a short bundle of microfilaments, precedes the development of a focal contact (Izzard and Lochner, 1980).

The leading lamellae of the marginal cells of an advancing epithelial sheet form close and focal contacts in much the same way as fibroblasts, and their subsequent fate is similar (Heath, 1982).

The contributions made by close and focal contacts to the processes involved in cell locomotion are not yet fully understood. Since close contacts are invariably formed near to the edge of an advancing leading lamella, often in front of any existing focal contacts, it is likely that they provide the adhesion required for the advance of the lamella (Izzard and Lochner, 1980). The fact that some cells which have only close contacts and lack focal contacts can move perfectly well (Couchman and Rees, 1979), whereas the movement of cells lacking close contacts has not been described, is in keeping with this possibility. But focal contacts presumably also have a part to play; they are invariably associated with a bundle of microfilaments which, as we have seen, are capable of contraction, at least in model systems (see Chapter 4, p. 68), and they are the regions of closest contact between cell and substratum. It is possible that the contraction of microfilament bundles associated with the focal contacts of the leading lamella could pull the rest of the cell towards these contacts and so generate forward movement (Abercrombie, 1980). Against this is the rather unexpected inverse relationship which exists between cell motility and the number of focal contacts and microfilament bundles possessed by the cell. In moving HeLa cells, for example, immunofluorescence has revealed that both actin and myosin are distributed diffusely throughout the cytoplasm, while in stationary cells of the same type most of these proteins are localised within microfilament bundles (Herman *et al.*, 1981). Similarly there is apparently an inverse relationship between the density of microfilament bundles and cell motility in human foreskin fibroblasts (Lewis *et al.*, 1982). Couchman and Rees (1979) have tried to clarify the relationship between motility and the different types of cell-substratum contacts in fibroblasts derived from explants of embryonic chick heart. The cells which migrate from these explants move rapidly during the first 48 hours in culture, but later their locomotory activity declines and they become almost stationary. During the motile phase, interference reflection microscopy reveals that the cells have conspicuous areas of close contact but lack focal contacts. Immunofluorescence and electron microscopy indicate that at this stage most of the actin is distributed diffusely as a meshwork of microfilaments, and that microfilament bundles are rare (Couchman and Rees, 1979). Later, focal contacts with their associated microfilament bundles are seen in some actively moving cells, but this type of contact becomes more common as the motility of the cells decreases. In addition, as the cells become almost stationary, contacts which have been called 'focal adhesions' appear (Couchman and Rees, 1979). These have an interference reflection image similar to that of focal contacts and are also associated with microfilament bundles, but they are larger and seem to be more stable than focal contacts (Couchman and Rees, 1979). It is interesting that, as the movement of the cells decreases, increasing amounts of fibronectin are associated with their surfaces, suggesting that it may contribute in some way to their immobilisation (Couchman and Rees, 1979). However, the presence of fibronectin has been shown to be necessary for the formation both of close contacts by mobile cells and of focal adhesions by stationary cells (Couchman *et al*., 1982). Kolega *et al*. (1982) have also examined the relationship between cell-substratum contacts and cell movement in a variety of cell types. They found that close contacts, but not focal contacts, were associated with rapid cell locomotion, and that increasing numbers of focal contacts were associated with decreasing cell movements.

Taken together, these results suggest that actively moving cells form close contacts in association with cytoplasmic meshworks of microfilaments and that focal contacts and focal adhesions with their microfilament bundles only develop as the cells become less motile. This must, however, be an oversimplification, since focal contacts with their microfilament bundles do occur in highly motile fibroblasts (Izzard and Lochner, 1976, 1980; Heath and Dunn, 1978). This apparent contradiction may result, at least in part, from the limits of resolution of the interference reflection microscope; the failure to detect focal contacts by this optical system is not proof of their absence. With the electron microscope Heath (1982) has found that, within what the interference reflection microscope reveals as a uniform close contact, small areas of the leading lamella approach to within 10 nm of the substratum – a separation characteristic of focal contacts. Heaysman and Pegrum (1982) have also found that close contacts can contain very small focal contacts, which can be visualised under the interference reflection microscope if the electrolyte concentration of the medium is changed so that the gap between cell and substratum is increased, thus abolishing the grey colour of the close contact. These authors believe that fibroblasts can only move over a substratum on which they can form focal contacts.

Hypotheses About the Mechanism of Cell Locomotion

From the evidence so far discussed it appears that cell locomotion can be considered, in simple terms, to result from the following sequence of events:

(i) protrusion of the cell margin free of the substratum to form a lamellipodium;
(ii) formation of adhesions between this lamellipodium and the substratum;
(iii) movement of the rest of the cell towards these adhesions.

As yet, we cannot define exactly the mechanisms by which these various activities are generated, neither do we know how they are co-ordinated. But in the light of the well-defined mechanisms which are now known to generate movement in striated muscle, a number of hypotheses have been put forward which attempt to explain movement by nonmuscle cells.

In striated muscle movement is generated by the development of a relative shearing force between filaments of actin and myosin and, as we have seen, many of the biochemical and structural properties of this interaction are shared by nonmuscle cells. It is possible that movement in these cells may also result from an active shearing mechanism in which sliding forces are developed between polarised actin filaments and some form of myosin assembly. Huxley (1973) has suggested that, in cells moving in culture, the actin of the microfilaments associated with cell-substratum contacts might provide one component of such a mechanism and cytoplasmic myosin the other. If the actin filaments were attached to the membrane with the same polarity as they have when attached to the Z-discs in muscle, then the cytoplasmic myosin would be expected to flow forwards over the attached filaments (Figure 6.8). Such a forward flow of cytoplasmic components could generate an internal pressure sufficient to cause the leading edge of the cell to advance. Fresh membrane attachment sites on to which actin could polymerise might be laid down at the newly protruded edge of the cell, leading to the formation of microfilaments, and if these became associated with cell-substratum contacts, cytoplasm could move forwards over them as before. It has, in fact, been established that the actin in the microfilament bundles associated with adhesion plaques (focal contacts) is organised with the polarity required by this model. When such microfilaments in PtK_2 cells are decorated with heavy meromyosin, the resulting 'arrowheads' point uniformly away from the cell membrane and towards the body of the cell (Sanger and Sanger, 1980).

Figure 6.8: Diagram summarising Huxley's proposed mechanism of cell locomotion. (Redrawn from Huxley, 1973.)

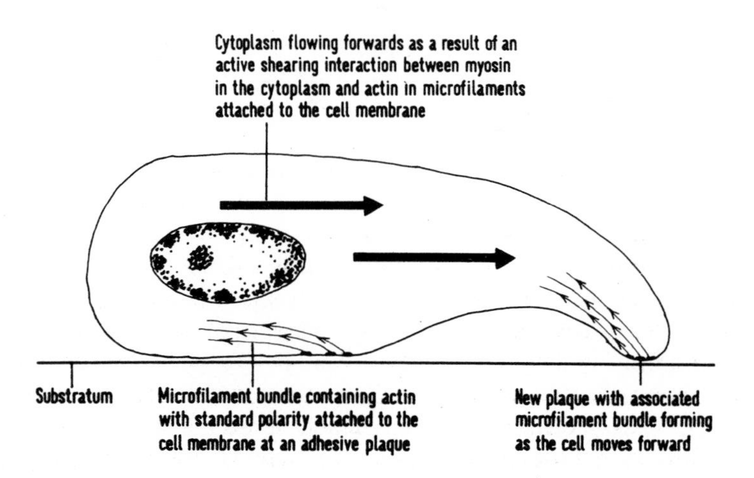

The polarity of actin in microfilament meshworks, rather than that in bundles, may be more relevant to cell movement. Ultrastructural studies using negative staining and immunocytochemical methods have demonstrated the presence of microfilament meshworks in the leading lamellae of a number of different fibroblastic cell types (Willingham *et al.*, 1981; Small *et al.*, 1978). Such meshworks are associated with close contacts (Heath and Dunn, 1978) and, as we have seen, during cell movement this type of contact often precedes the formation of focal contacts and their associated microfilament bundles. When the microfilaments in these meshworks are decorated with HMM the arrowheads all point away from the direction of cell movement and towards the body of the cell (Small *et al.*, 1978). Given the polarity of the actin in these meshworks their microfilaments could, if associated with close contacts, contribute to cell locomotion in the way that Huxley has suggested.

Some modifications to Huxley's scheme have been put forward by Small *et al.* (1978); they assume that the microfilament meshwork is associated with cell-substratum contacts, and suggest that the advance of the leading lamella involves a shearing between the actin in this meshwork and cytoplasmic myosin, which, they propose, would bring monomeric actin and membrane precursor components within the cytoplasm to the tip of the advancing edge for the polymerisation of actin filaments and the production of new surface membrane. In this case the actin, rather than polymerising inwards from the cell membrane (as envisaged by Huxley), would polymerise in a forward direction and so advance the leading edge of the meshwork. Such polymerisation implies that actin monomers are being added to the 'barbed' ends of actin filaments, and this is consistent with the observation that isolated actin filaments do elongate preferentially in this way (see Chapter 4, p. 71). If the newly formed meshwork at the advancing edge of the cell became associated with cell-substratum contacts, the shearing process could continue, so further advancing the cell. Ruffling of the leading lamella might be the result of this same shearing mechanism operating in the absence of an association between meshwork microfilaments and cell-substratum contacts (Small *et al.*, 1978).

In this scheme the protrusion of the leading edge is due primarily to the polymerisation of actin rather than to an actin-myosin interaction. A similar proposal was made by Abercrombie (1980) and is supported by immunofluorescence observations which reveal a relative deficiency of myosin in the leading lamellae of fibroblasts (Herman *et al.*, 1981).

In at least one system, protrusive activity has been shown to result from the explosive polymerisation of actin. Shortly after making contact with the jelly that surrounds the eggs, the sperm of many marine invertebrates produce a long acrosomal process. In the echinoderm, *Thyone*, for example, this process develops within 30 seconds and may be up to 90 μm long. Its formation is the result of the polymerisation of a cytoplasmic pool of G-actin to F-actin which forms a microfilament bundle running the length of the acrosomal process (Tilney *et al.*, 1973). Potentially, the substantial amount of G-actin present in cells in tissue culture (see Chapter 4, p. 59) could provide a suitable pool of actin for polymerisation during protrusive activity; but as yet there is no direct evidence that protrusion does result from such polymerisation.

The proposals of both Huxley (1973) and Small *et al.* (1978) suggest that protrusive activity is associated with a forward flow inside the cell of cytoplasmic constituents and, perhaps, of membrane precursors. There is no doubt that such a flow must occur. As a fibroblast advances its focal contacts and their microfilament bundles remain stationary; as a result, the material associated with these structures, and any similar material concerned in movement, is progressively lost from the front of the cell. There must, therefore, be a compensating forward flow of equivalent material to replace it (Abercrombie, 1980). Much of this material is likely to be actin needed to form both the microfilament bundles associated with focal contacts and the microfilament meshwork associated with close contacts and present in the cortical cytoplasm. It may well be that this flows forward as G-actin as suggested by Small *et al.* (1978), but this has not been demonstrated. In fact, it is usually impossible to detect the forward flow of cytoplasmic constituents in moving fibroblasts, but such a flow must exist and certainly merits further investigation.

Surprisingly, in view of the foregoing and in the context of forward locomotion, structures (or artificial markers) at or near the front of cells move predominantly *backwards* as the cell moves forwards. The rearward movement of ruffles over the leading lamella towards the nucleus has already been discussed, but this is not an isolated example, and blebs and microvilli present on the leading lamella behave in the same manner (Harris, 1973a). Similarly, small fragments of debris or particles which become attached to the leading lamella move backwards, relative to the substratum, as fibroblasts or epithelial cells advance (Abercrombie *et al.*, 1970c; Di Pasquale, 1975b). Abercrombie *et al.* (1970c) thought that this movement could be explained if the cell membrane to which these structures or markers were attached flowed bodily backwards from the front edge of the leading lamella towards the nucleus in the course of cell movement. They suggested that such a rearward flow might be generated if membrane materials were continuously being assembled and inserted into the leading lamella close to its advancing edge and continuously being disassembled and removed from the surface somewhere closer to the nucleus.

To conform with the rates of turnover known to exist for certain membrane components this suggestion requires that the cycle be completed by a forward flow of the disassembled membrane materials towards the site of assembly for reutilisation (Figure 6.9). Such a cycle, in addition to explaining the backward movement of structures on the leading lamella, could potentially account for the protrusion and withdrawal of lamellipodia seen at the edge of an advancing leading lamella.

Protrusion would result if the rate of assembly exceeded the rate of rearward flow; if rearward flow exceeded assembly, then withdrawal of the lamellipodia would result (Abercrombie *et al.*, 1970c). The formation of ruffles could also be explained by the same mechanism. The rearward flow of membrane would be expected to happen predominantly on the upper surface of the cell, since the cell-substratum adhesions would tend to block it on the lower surface. Consequently, the surface pressure on the underside of a lamellipodium would be higher than on its upper surface, so the lamellipodium would tend to curl upwards away from the substratum, and, if the pressure gradient became sufficiently steep, a ruffle might form (Abercrombie *et al.*, 1970c).

Figure 6.9: Diagram summarising the explanation proposed by Abercrombie *et al.* (1970c) for the cycle of protrusion and withdrawal of lamellipodia. (Redrawn from Harris, 1973a.)

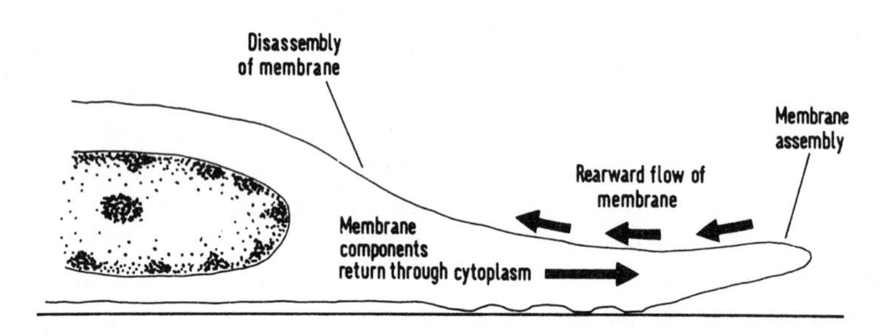

In an attempt to confirm the possibility that cell locomotion involves the assembly and subsequent backward flow of membrane, Abercrombie *et al.* (1972) directly labelled components of the cell membrane in living fibroblasts, using the plant lectin, Concanavalin A, which binds specifically to certain carbohydrates in the cell surface. The labelling was done in the cold so that the cells were not moving, and they were then transferred to lectin-free medium at 37°C to allow movement to resume. After allowing the cells to move for varying lengths of time, the cultures were fixed and the lectin treated to render it visible in the electron microscope. The experiment showed that, as the cells moved, the labelled membrane gradually retreated from the advancing edge of the leading lamella and was replaced by unlabelled membrane, a result entirely consistent with the mechanism suggested by Abercrombie and his colleagues. Their suggestion was extended and elaborated by Harris (1973a). He argued that, if a backward flow of cell membrane could exert a force on particles attached to the upper surface of the leading lamella, then the same force, if applied to the substratum by the lower surface of the lamella, could contribute to the forward motion of the cell. Evidence that the lower surface of the lamella could exert such a force was provided by the observation that particles which become attached to this surface move backwards during cell locomotion in a similar way to particles attached to the upper surface (Harris and Dunn, 1972). This carries the implication that there is also a rearward flow of membrane on the lower surface of the leading lamella, and the application of the force generated by this flow to the fixed substratum, rather than to a movable particle, would tend to move the cell forward.

Thus the assembly and backward flow of membrane could account for two of the observed features of cell locomotion: the forward protrusion of the leading lamella and the exertion of a rearward force on the substratum. Harris felt, however, that the membrane would be too flexible to flow backwards under the pressure of its own assembly as suggested by Abercrombie *et al.* (1970c). Instead, he suggested that the microfilaments lying in the cortical cytoplasm within the leading lamella might be responsible for the backward flow either by pulling directly on the membrane or by interacting with cytoplasmic myosin.

Unfortunately, several of the proposals discussed above have been made less acceptable by more recent work. It is now known that a number of cell membrane components can be redistributed, in the plane of the surface membrane, into numerous small patches or a single large cap by treatments which form cross-links between the components. Such redistributions occur, for example, when cells are treated with ligands, such as Concanavalin A, which bind to specific carbohydrates in the membrane, and also when they are exposed to antibodies against cell membrane antigens. The redistribution of membrane components in these cases apparently results only from the movement of the cross-linked components and there is no movement of the membrane as a whole (see Nicolson, 1979; Weatherbee, 1981). It has consequently been pointed out that the backward movement of particles, or Concanavalin A receptors, seen in cell locomotion could be the result of similar cross-link induced redistributions, and does not prove that there is backward flow of the membrane itself. If this is the case, the cycle of membrane assembly and disassembly, envisaged by Abercrombie *et al.* (1970c) as an explanation of this flow, may not exist and hence could not be responsible for the protrusion of the leading lamella.

Even if there is no backward movement of the membrane as a whole during cell locomotion, there is no doubt that membranous protrusions in the form of ruffles and blebs do move backwards as cells advance. Dunn (1980) has proposed that both the movement of these structures and the protrusion of the leading lamella could be generated by a cytoplasmic contractile meshwork which, he postulates, is present throughout the cell. The microtrabecular lattice (see Chapter 4, p. 78) might provide the structural basis for such a meshwork, since there is some evidence that it contains actin (Webster *et al.*, 1978). Alternatively, the network of microfilaments described in the leading lamella of fibroblasts by Small *et al.* (1978) might be suitable if it is present throughout the cell. Another possibility is that the contractile actions of different microfilaments are co-ordinated somehow to produce a functional meshwork. Schliwa and Blerkum (1981) have described a new actin-free class of cytoplasmic filaments 2-3 nm in diameter which form end-to-side contacts with the various components of the cytoskeleton, apparently linking them together. Such filaments could co-ordinate the actions of microfilaments in different parts of the cell, and so establish a contractile network. Whatever its structural basis, Dunn (1980) proposes that, as a cell moves, this network undergoes continuous contraction from the cell margin towards the nucleus. If the contraction is to be continuous, the meshwork must continuously disassemble in some more central region of the cell, possibly near the nucleus, and then be transported back to the marginal region to reassemble and participate again in the contraction. This cycle is similar to that originally proposed for the cell membrane (Abercrombie *et al.*, 1970c) and again involves a forward flow of material within the cytoplasm, but the emphasis has shifted from membrane to cytoplasmic components.

Evidence that there is a rearward contraction of the cytoplasm during cell locomotion comes from the observation that, under suitable conditions, the cytoplasm of some glycerinated cell models can be induced to contract towards the nucleus (Isenberg *et al.*, 1976). In addition, in moving fibroblasts, cytoplasmic structures known as 'arcs' can be seen moving backwards through the cytoplasm as the cell proceeds forwards; arcs are visible with phase contrast optics in the leading lamellae of fibroblasts from several different sources (Heath and Dunn, 1978; Heath, 1981), appearing as slightly curved phase-dark lines up to 60 μm in length and about 2 μm wide (Figure 6.10). They are first seen close to the margin of the leading lamella and move backwards through the lamella towards the nucleus at 1.5-3.0 μm per minute as the cells move forwards. On reaching the perinuclear region they disappear (Heath, 1981). Electron microscopy shows that they consist of closely packed microfilaments oriented parallel to the margin of the leading lamella (Heath, 1981). Dunn (1980) considers that the movement of these structures is indicative of the movement of the postulated meshwork generated by its rearward contraction.

Figure 6.10: Phase contrast photomicrograph of a living chick heart fibroblast showing an arc. (Courtesy of Dr J. Heath.)

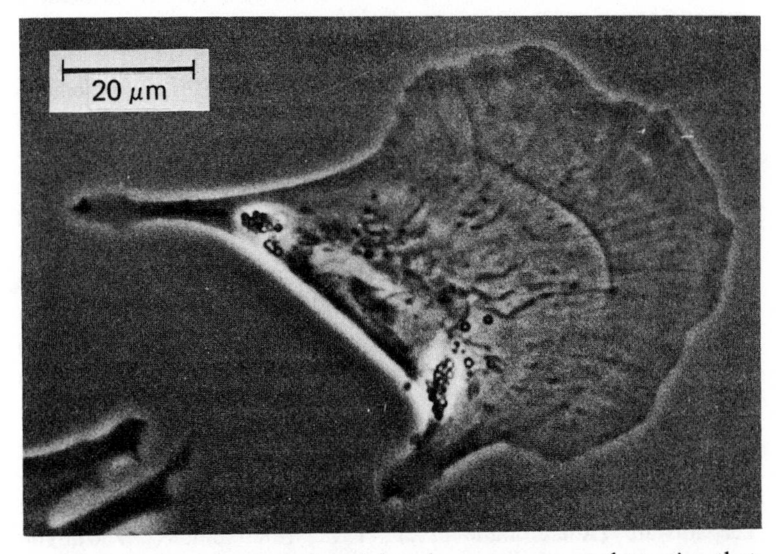

There is also some experimental evidence to support the notion that contraction of such a meshwork might explain the protrusive activity of the leading lamella. When the elongated, narrow 'tail' at the rear of a moving fibroblast is detached from the substratum with a micromanipulator needle, it retracts towards the body of the cell and the leading lamella then invariably displays a wave of increased protrusive activity (Chen, 1981a; Dunn, 1980). Since tail retraction can be partially restricted by metabolic inhibitors, it must, at least in part, be due to an active contraction (Chen, 1981b). The fact that the contraction in the rear of the cell is followed immediately by increased protrusive activity at the advancing edge raises the possibility that the protrusive activity seen during normal locomotion could be the outcome of a continuous active contraction of the cytoplasm. Dunn (1980) has, therefore, postulated that protrusion is generated by the reassembly of the proposed cytoplasmic meshwork at the margin of the leading lamella. He considers that the surface membrane of the cell plays a passive role in this process, and that protrusion is part of a continuous cycle of network contraction, disassembly, transport to the periphery, reassembly and contraction. His concept is supported by the observation that, following the contraction induced by tail detachment, there is a delay of 10-20 seconds before the increase in protrusive activity at the front of the cell reaches its peak. This, it is thought, might represent the time required for the components of the network to be disassembled and transported forward to the edge of the leading lamella for reassembly (Dunn, 1980).

One virtue of this hypothesis of continuous contraction is that it could provide a simple explanation of the backward movement of ruffles and other protrusions over the surface of the leading lamella during cell locomotion. All that is required is that the cytoplasm within the protrusions should be integrated mechanically with the meshwork. If this prerequisite is fulfilled, the protrusions must inevitably move backwards as the meshwork contracts towards the nucleus. However, it is not so easy to explain the backward movement of particles and other markers attached to the membrane; such movement would seem to require either that the membrane is linked to the meshwork, or that the markers themselves somehow establish links with the meshwork across the membrane.

It has been suggested that during cell locomotion the microfilament bundles associated with the focal contacts of the leading lamella contract, so exerting a force on the substratum and pulling the rest of the cell forward (Abercrombie *et al.*, 1971; Abercrombie, 1980). The continuous contraction hypothesis potentially provides an explanation of the formation of these bundles and for the generation of the force. Fleischer and Wohlfarth-Botterman (1975) have demonstrated that microfilament bundles are formed in the cytoplasm of the slime mould *Physarum* when tension is generated as a result of isometric contraction. Conversely, when the tension in a fibroblast 'tail' is released by detachment from the substratum, the microfilament bundles previously present in the tail become rearranged into a meshwork of microfilaments (Chen, 1981b). Abercrombie *et al.* (1977) suggested that microfilament bundles might form in fibroblasts when mechanical stress was developed in part of the cytoplasmic microfilament meshwork. If there is a cytoplasmic meshwork continuously contracting towards the nucleus, the formation of a focal contact might effectively anchor part of the meshwork to the substratum and so prevent it from moving towards the nucleus. This could cause the rapid development of stress in the meshwork between the focal contact and the perinuclear region and result in the formation of the microfilament bundle that is found running obliquely upwards from a focal contact towards the nucleus (Dunn, 1980). In fact, the microfilament bundles associated with focal contacts shorten as the cell moves forward and therefore their contraction is not strictly isometric; the speed at which they shorten is about the same as the speed of forward motion of the cell, approximately 1.0 μm per minute, which is considerably slower than the estimated speed of the postulated network contraction. Based on the velocity of arcs, this is about 2.0-3.0 μm per minute (Dunn, 1980; Heath, 1981). It may be that, rather than effectively anchoring part of the network to the substratum, the formation of a focal contact reduces the speed of contraction of a portion of the network, and this may generate enough stress to cause the formation of a microfilament bundle (Dunn, 1980).

As we have seen, some cells which do not have focal contacts are still capable of locomotion. In these cells the force necessary for locomotion must presumably be exerted on the substratum via the close contacts, which are not usually associated with microfilament bundles (Heath and Dunn, 1978). Dunn (1980) speculates that the proposed contractile meshwork could exert this force, which must be too small to cause the formation of microfilament bundles, perhaps because the meshwork may be connected less directly to the substratum at close contacts than at focal contacts.

The continuous contraction mechanism is an elegant and ingenious hypothesis, but must, for the present, be open to some doubt. There is so far only indirect evidence for the continuous rearward contraction of a (contractile) meshwork; further, a continuous cycle of assembly and disassembly of the proposed network has not been demonstrated. In spite of these uncertainties, the hypothesis has the potential to account for a number of the phenomena associated with cell locomotion, and it will certainly stimulate further research.

Any attempted explanation of cell locomotion must suggest a mechanism for confining protrusive activity to a limited part of the cell margin so that the movement of the cell is polarised. Unless this happens, protrusive activity will occur all round the edge of the cell, and directed movement will be impossible. Experiments with drugs which disrupt microtubules lead to the conclusion that these structures are involved both in the establishment and in the maintenance of polarisation (see p. 108), but little is known about the mechanisms involved. We have seen that locomotion must involve a forward flow of material in the cytoplasm, although there are differing views about the exact nature of this material. Microtubules could polarise cell locomotion by directing this flow towards a limited part of the cell margin, thus determining the position of the protrusive activity.

Albrecht-Buehler (1977a) has suggested that MTOCs (see p. 94) may be important in determining the direction of locomotion of cells in culture. He has observed the locomotion of 3T3 cells (and those of some other cell lines) after they have divided, and found that, in the majority of cases, sister cells appear to follow paths that are identical or which exhibit mirror image symmetry (Albrecht-Buehler, 1977b). Indeed, in some instances the path followed by an ancestral cell seems to be repeated by its descendants for up to three generations (Albrecht-Buehler, 1977c).

These surprising findings suggest that the direction of cell locomotion could be predetermined, and raise the possibility that a cell component which is duplicated and shared between the cells during mitosis may be responsible (Albrecht-Buehler, 1977a). In an immunofluorescence study of 3T3 cells the MTOC was found almost invariably to be orientated parallel to the direction of movement of the cell; this observation by Albrecht-Buehler (1977a), and the implication that the MTOC may be fundamentally important in deciding the direction in which a cell moves, is obviously of potential importance. But some recent work has cast doubt on this. A detailed analysis of the movements of human fibroblasts has revealed that, although the paths of sister cells appear to be similar, they do not in fact exhibit rigorous symmetry (Levinstone *et al.*, 1983). However, the work of Gotlieb *et al.* (1981), and that of Kupfer *et al.* (1982), based on immunofluorescence labelling of the MTOC in fibroblasts and endothelial cells, has shown that when the cells are stationary the MTOCs are randomly positioned within the cytoplasm, but after the cells begin to move the MTOC is almost always found between the nucleus and the advancing edge of the cell. What is not clear is whether this relocation of the MTOC initiates locomotion in a particular direction or is merely a consequence of the onset of movement.

It is obvious that we still have a lot to learn about the mechanisms underlying cell locomotion, but it is certain that nonmuscle cells contain proteins which are biochemically and structurally related to those which are known to be responsible for the contraction of striated muscle. It seems highly probable, therefore, that movement is generated, in nonmuscle cells as in muscle fibres, by essentially similar mechanisms. In fact, as Huxley (1973) suggested, it is likely that these mechanisms developed very early in evolution, and that the interactions essential to them have been conserved ever since. The contraction of striated muscle may represent no more than a highly specialised and elaborate example of a basic mechanism present in nonmuscle cells and responsible for their locomotion.

References

M. Abercrombie (1980) 'The Crawling Movement of Metazoan Cells', *Proc. Roy. Soc. B*, vol. 207, p. 129

— and G.A. Dunn (1975) 'Adhesion of Fibroblasts to Substratum During Contact Inhibition Observed by Interference Reflection Microscopy', *Exp. Cell Res.*, vol. 92, p. 57

—, G.A. Dunn and J.P. Heath (1977) 'The Shape and Movement of Fibroblasts in Culture', in J.W. Lash and M.M. Burger (eds.), *Cell and Tissue Interactions* (Raven Press, New York), pp. 57-70

—, J.E.M. Heaysman and S.M. Pegrum (1970a) 'The Locomotion of Fibroblasts in Culture. I. Movements of the Leading Edge', *Exp. Cell Res.*, vol. 59, p. 393

—, J.E.M. Heaysman and S.M. Pegrum (1970b) 'The Locomotion of Fibroblasts in Culture. II. Ruffling', *Exp. Cell Res.*, vol. 60, p. 437

—, J.E.M. Heaysman and S.M. Pegrum (1970c) 'The Locomotion of Fibroblasts in Culture. III. Movements of Particles on the Dorsal Surface of the Leading Lamella', *Exp. Cell Res.*, vol. 62, p. 389

—, J.E.M. Heaysman and S.M. Pegrum (1971) 'The Locomotion of Fibroblasts in Culture. IV. Electron Microscopy of the Leading Lamella', *Exp. Cell Res.*, vol. 67, p. 359

—, J.E.M. Heaysman and S.M. Pegrum (1972) 'The Locomotion of Fibroblasts in Culture. V. Surface Marking with Concanavalin A', *Exp. Cell Res.*, vol. 73, p. 536

G. Albrecht-Buehler (1977a) 'Phagokinetic Tracks of 3T3 Cells: Parallels Between the Orientation of Track Segments and of Cellular Structures Which Contain Actin or Tubulin', *Cell*, vol. 12, p. 333

— (1977b) 'Daughter 3T3 Cells: Are They Mirror Images of Each Other?', *J. Cell Biol.*, vol. 72, p. 595

— (1977c) 'The Phagokinetic Tracks of 3T3 Cells', *Cell*, vol. 11, p. 395

R.A. Badley, C.W. Lloyd, A. Woods, L. Carruthers, C. Allcock and D.A. Rees (1978) 'Mechanisms of Cellular Adhesion. III. Preparation and Preliminary Characterisation of Adhesions', *Exp. Cell Res.*, vol. 117, p. 231

A.D. Bershadsky, V.I. Gelfand, T.M. Svitkina and I.S. Tint (1980) 'Destruction of Microfilament Bundles in Mouse Embryo Fibroblasts Treated with Inhibitors of Energy Metabolism', *Exp. Cell Res.*, vol. 127, p. 421

Z.L. Bliokh, V. Domnina, O.Y. Ivanova, O.Y. Pletjushkina, T.M. Svitkina, V.A. Smolyaninov, J.M. Vasiliev and I.M. Gelfand (1980) 'Spreading of Fibroblasts in Medium Containing Cytochalasin B: Formation of Lamellar Cytoplasm as a Combination of Several Functionally Different Processes', *Proc. Nat. Acad. Sci.*, vol. 77, p. 5919

E.E. Bragina, J.M. Vasiliev and I.M. Gelfand (1976) 'Formation of Bundles of Microfilaments During Spreading of Fibroblasts on the Substrate', *Exp. Cell Res.*, vol. 97, p. 241

K. Burridge and J.R. Feramisco (1980) 'Microinjection and Localisation of a 130K Protein in Living Fibroblasts: A Relationship to Actin and Fibronectin', *Cell.*, vol. 19, p. 587

— and L. McCullough (1980) 'The Association of α-Actinin with the Plasma Membrane', *J. Supramol. Struct.*, vol. 13, p. 53

W.-T. Chen (1981a) 'Surface Changes During Retraction-Induced Spreading', *J. Cell Sci.*, vol. 49, p. 1
— (1981b) 'Mechanism of Retraction of the Trailing Edge During Fibroblast Movement', *J. Cell Biol.*, vol. 90, p. 187
— and S.J. Singer (1980) 'Fibronectin is not Present in the Focal Adhesions Formed Between Normal Cultured Fibroblasts and their Substrata', *Proc. Nat. Acad. Sci.*, vol. 77, p. 7318
J.R. Couchman and D.A. Rees (1979) 'The Behaviour of Fibroblasts Migrating from Chick Heart Explants: Changes in Adhesion, Locomotion and Growth and in the Distribution of Actomyosin and Fibronectin', *J. Cell Sci.*, vol. 39, p. 149
—, D.A. Rees, M.R. Green and C.G. Smith (1982) 'Fibronectin has a Dual Role in Locomotion and Anchorage of Primary Chick Fibroblasts and can Promote Entry into the Division Cycle', *J. Cell Biol.*, vol. 93, p. 402
A. Di Pasquale (1975a) 'Locomotory Activity of Epithelial Cells in Culture', *Exp. Cell* Res., vol. 94, p. 191
— (1975b) 'Locomotion of Epithelial Cells. Factors Involved in Extension of the Leading Edge', *Exp. Cell Res.*, vol. 95, p. 425
G.A. Dunn (1980) 'Mechanisms of Fibroblast Locomotion', in A.S.G. Curtis and J.D. Pitts (eds.), *Cell Adhesion and Motility* (Third Symposium British Society for Cell Biology, Cambridge University Press), pp. 409-23
C.A. Erickson and J.P. Trinkaus (1976) 'Microvilli and Blebs as Sources of Reserve Surface Material During Cell Spreading', *Exp. Cell Res.*, vol. 99, p. 375
M. Fleischer and K.E. Wohlfarth-Botterman (1975) 'Correlation Between Tension Force Generation, Fibrillogenesis and Ultrastructure of Cytoplasmic Actomyosin During Isometric and Isotonic Contraction of Protoplasmic Strands', *Cytobiologie*, vol. 10, p. 339
M.H. Gail and C.W. Boone (1970) 'The Locomotion of Mouse Fibroblasts in Tissue Culture', *Biophys. J.*, vol. 10, p. 980
— and C.W. Boone (1971) 'Effect of Colcemid on Fibroblast Motility', *Exp. Cell Res.*, vol. 65, p. 221
B. Geiger (1979) 'A 130K Protein from Chicken Gizzard: Its Localisation at the Termini of Microfilament Bundles in Cultured Chicken Cells', *Cell*, vol. 18, p. 193
R.D. Goldman and D. Knipe (1973) 'Functions of Cytoplasmic Fibres in Non-Muscle Cell Motility', *Cold Spring Harbor Symp. Quant. Biol.*, vol. 37, p. 523
—, J.A. Schloss and J.M. Starger (1976) 'Organisational Changes of Actin-like Microfilaments During Animal Cell Movement', in R. Goldman, T. Pollard and J.L. Rosenbaum (eds.), *Cell Motility* Book A (Cold Spring Harbor), pp. 217-45
W.E. Gordon III and A: Bushnell (1979) 'Immunofluorescent and Ultrastructural Studies of Polygonal Microfilament Networks in Respreading Non-Muscle Cells', *Exp. Cell Res.*, vol. 120, p. 335
A.I. Gotlieb, L.M. May, L. Subrahmanyan and V.I. Kalnins (1981) 'Distribution of Microtubule Organising Centres in Migrating Sheets of Endothelial Cells', *J. Cell Biol.*, vol. 91, p. 589
F. Grinnell (1982) 'Migration of Human Neutrophils in Hydrated Collagen Lattices', *J. Cell Sci.*, vol. 58, p. 95
A.K. Harris (1973a) 'Cell Surface Movements Related to Cell Locomotion', in R. Porter and D.W. Fitzsimons (eds.), *Locomotion of Tissue Cells,* Ciba Foundation Symposium No. 14 (New Series), (Ass. Scient. Publishers, Amsterdam), pp. 3-20
— (1973b) 'Location of Cellular Adhesions to Solid Substrata', *Dev. Biol.*, vol.

35, p. 97
— and G. Dunn (1972) 'Centripetal Transport of Attached Particles on Both Surfaces of Moving Fibroblasts', *Exp. Cell Res.*, vol. 73, p. 519
J.P. Heath (1981) 'Arcs: Curved Microfilament Bundles Beneath the Dorsal Surface of the Leading Lamella of Moving Chick Embryo Fibroblasts', *Cell Biol. Int. Reps.*, vol. 5, p. 975
— (1982) 'Adhesions to Substratum and Locomotory Behaviour of Fibroblastic and Epithelial Cells in Culture', in R. Bellairs, A. Curtis and G. Dunn (eds.), *Cell Behaviour* (Cambridge University Press), pp. 77-108
— and G.A. Dunn (1978) 'Cell-to-Substratum Contacts of Chick Fibroblasts and Their Relation to the Microfilament System. A Correlated Interference-Reflexion and High Voltage Electron Microscope Study', *J. Cell Sci.*, vol. 29, p. 197
J.E.M. Heaysman and S.M. Pegrum (1982) 'Early Cell Contacts in Culture', in R. Bellairs, A. Curtis and G. Dunn (eds.), *Cell Behaviour* (Cambridge University Press), pp. 49-76
I.M. Herman, N.J. Crisona and T.D. Pollard (1981) 'Relation Between Cell Activity and the Distribution of Cytoplasmic Actin and Myosin', *J. Cell Biol.*, vol. 90, p. 84
H.E. Huxley (1973) 'Muscular Contraction and Cell Motility', *Nature*, vol. 243, p. 445
R.O. Hynes and A.T. Destree (1978) 'Relationship Between Fibronectin (LETS Protein) and Actin', *Cell*, vol. 15, p. 875
V.M. Ingram (1969) 'A Side View of Moving Fibroblasts', *Nature*, vol. 272, p. 641
G. Isenberg, P.C. Rathke, N. Hulsmann, N.W. Franke and K.E. Wohlfarth-Botterman (1976) 'Cytoplasmic Actomyosin Fibrils in Tissue Culture Cells. Direct Proof of Contractility by Visualization of ATP-Induced Contraction in Fibrils Isolated by Laser Microbeam Dissection', *Cell Tiss. Res.*, vol. 166, p. 427
C.S. Izzard and L.R. Lochner (1976) 'Cell-to-Substrate Contacts in Living Fibroblasts. An Interference-Reflexion Study with an Evaluation of the Technique', *J. Cell Sci.*, vol. 21, p. 129
— and L.R. Lochner (1980) 'Formation of Cell-to-Substrate Contacts During Fibroblast Motility: An Interference-Reflexion Study', *J. Cell Sci.*, vol. 42, p. 81
B.M. Jockusch and G. Isenberg (1981) 'Interaction of α-Actinin and Vinculin with Actin: Opposite Effects on Filament Network Formation', *Proc. Nat. Acad. Sci.*, vol. 78, p. 3005
J. Kolega, M.S. Shure, W.-T. Chen and N.D. Young (1982) 'Rapid Cellular Translocation is Related to Close Contacts formed Between Various Cultured Cells and Their Substrata', *J. Cell Sci.*, vol. 54, p. 23
A. Kupfer, D. Lonvard and S.J. Singer (1982) 'Polarisation of the Golgi Apparatus and the Microtubule Organising Centre in Cultured Fibroblasts at the Edge of an Experimental Wound', *Proc. Nat. Acad. Sci.*, vol. 79, p. 2603
E. Lazarides (1976a) 'Actin, α-Actinin and Tropomyosin Interaction in the Structural Organisation of Actin Filaments in Non-Muscle Cells', *J. Cell Biol.*, vol. 68, p. 202
— (1976b) 'Aspects of the Structural Organisation of Actin Filaments in Tissue Culture Cells', in R. Goldman, T. Pollard and J.L. Rosenbaum (eds.), *Cell Motility* Book A (Cold Spring Harbor), pp. 347-60
— and K. Burridge (1975) 'α-Actinin: Immunofluorescent Localisation of a Muscle Structural Protein in Non-Muscle Cells', *Cell*, vol. 6, p. 289

D. Levinstone, M. Eden and E. Bell (1983) 'Similarity of Sister-Cell Trajectories in Fibroblast Clones', *J. Cell Sci.*, vol. 59, p. 105
L. Lewis, J.-M. Verna, D. Levinstone, S. Sher, L. Marek and E. Bell (1982) 'The Relationship of Fibroblast Translocations to Cell Morphology and Stress Fibre Density', *J. Cell Sci.*, vol. 53, p. 21
L. Lochner and C.S. Izzard (1973) 'Dynamic Aspects of Cell-Substrate Contact in Fibroblast Motility', *J. Cell Biol.*, vol. 59, p. 199a
G.L. Nicolson (1979) 'Topographic Display of Cell Surface Components and Their Role in Transmembrane Signalling', *Curr. Topics Dev. Biol.*, vol. 13, p. 305
M. Osborn and K. Weber (1976) 'Tubulin-Specific Antibody and the Expression of Microtubules in 3T3 Cells After Attachment to a Substratum. Further Evidence for the Polar Growth of Cytoplasmic Microtubules *in vivo*', *Exp. Cell Res.*, vol. 103, p. 331
J.M. Sanger and J.W. Sanger (1980) 'Banding and Polarity of Actin Filaments in Interphase and Cleaving Cells', *J. Cell Biol.*, vol. 86, p. 568
M. Schliwa and J. Blerkum (1981) 'Structural Interaction of Cytoskeletal Components', *J. Cell Biol.*, vol. 90, p. 222
S.L. Schor (1980) 'Cell Proliferation and Migration on Collagen Substrata *in vitro*', *J. Cell Sci.*, vol. 41, p. 159
J.V. Small, G. Isenberg and J.E. Celis (1978) 'Polarity of Actin at the Leading Edge of Cultured Cells', *Nature*, vol. 272, p. 638
L.G. Tilney, S. Hatano, H. Ishikawa and M.S. Mooseker (1973) 'The Polymerisation of Actin: Its Role in the Generation of the Acrosomal Process of Certain Echinoderm Sperm', *J. Cell Biol.*, vol. 59, p. 109
J.M. Vasiliev and I.M. Gelfand (1981) *Neoplastic and Normal Cells in Culture* (Cambridge University Press), p. 392
J.A. Weatherbee (1981) 'Membranes and Cell Movement: Interactions of Membranes With the Proteins of the Cytoskeleton', *Int. Rev. Cytol.*, Suppl. 12, p. 113
R.E. Webster, D. Henderson, M. Osborn and K. Weber (1978) 'Three Dimensional Electron Microscopical Visualisation of the Cytoskeleton of Animal Cells: Immunoferritin Identification of Actin and Tubulin Containing Structures', *Proc. Nat. Acad. Sci.*, vol. 75, p. 5511
J. Wehland, M. Osborn and K. Weber (1979) 'Cell-to-Substratum Contacts in Living Cells: A Direct Correlation Between Interference-Reflexion and Indirect Immunofluorescence Microscopy Using Antibodies Against Actin and α-Actinin', *J. Cell Sci.*, vol. 37, p. 257
N.K. Wessells (1982) 'Axon Elongation: A Special Case of Cell Locomotion', in R. Bellairs, A. Curtis and G. Dunn (eds.), *Cell Behaviour* (Cambridge University Press), pp. 225-46
M.C. Willingham, S.S. Yamada, P.J.A. Davies, A.V. Rutherford, M.G. Gallo and I. Pastan (1981) 'Intracellular Localization of Actin in Cultured Fibroblasts by Electron Microscopic Immunocytochemistry', *J. Histochem. Cytochem.*, vol. 29, p. 17
K.M. Yamada and K. Olden (1978) 'Fibronectins – Adhesive Glycoproteins of Cell Surface and Blood', *Nature*, vol. 275, p. 179

7 THE SOCIAL BEHAVIOUR OF CELLS IN CULTURE

The locomotion of cells in culture can be influenced by a number of factors, including the nature of the culture medium and of the substratum. The behaviour of individual cells can also be influenced by their neighbours, and hence a population of cells may be said to exhibit social behaviour (Abercrombie and Heaysman, 1953). In this chapter we consider two examples of this, known respectively as contact inhibition and contact-induced spreading.

Contact Inhibition

Normal cells cultured on a plane substratum usually arrange themselves approximately into a layer one cell thick. The corollary of this observation is that the cells do not move over each other's surface to form a multilayered organisation. In a series of experiments (now regarded as classic) Michael Abercrombie and Joan Heaysman set out to establish why this should be so. They investigated the interactions between fibroblasts migrating out from explants of chick embryo heart on to a glass substratum; to maximise interactions between the cells, they placed two explants about 1 mm apart so that the outgrowths from them would collide. They analysed the movement and distribution of the cells before and after the outgrowths met and found that, soon after colliding, the cells showed a marked change in their behaviour. Their speed of locomotion decreased sharply, their direction of movement became approximately random (instead of predominantly outwards from the explant), and the population density in the space between the explants became almost stable. Consequently, the cells became virtually stationary and remained more or less as a monolayer on the substratum (Abercrombie and Heaysman, 1954). They interpreted these observations as indicating that fibroblasts are somehow inhibited from moving over each other. Since the cells showed no tendency to reduce their speed of locomotion until after the two outgrowths had collided, they concluded that this inhibition of movement was a result of contact between the cells, and named the phenomenon 'contact inhibition' (Abercrombie and Heaysman, 1954). It has since been defined as 'the prohibition, when contact between cells occurred, of continued movement such as would carry one cell over the surface of another' (Abercrombie, 1970). Some confusion has arisen because the term contact inhibition has also been used to describe the reduction in the frequency of cell division which occurs in dense cultures of many types of cell. This phenomenon is probably better referred to as 'density-dependent inhibition of mitosis' (Stoker and Rubin, 1967), and the term 'contact inhibition' should be used only in its original sense to refer to a directional prohibition of movement resulting from cell contact.

Abercrombie and Heaysman (1954) suggested that one explanation of contact inhibition might be that the surface activity involved in fibroblast locomotion is paralysed in those parts of its surface which are in contact with other cells, which would effectively stop further movement in the direction that led to the contact. This idea was supported by the results of a previous study which had demonstrated an inverse relationship between the velocity of fibroblasts and the number of other cells they were in contact with. Thus while isolated fibroblasts lacking contacts with other cells had a mean velocity of about 1.4 μm per minute, fibroblasts in contact with five other cells had a mean velocity of only about 0.7 μm per minute (Abercrombie and Heaysman, 1953). Confirmation that locomotory surface activity is indeed inhibited in regions of cell contact was provided by time-lapse films of colliding chick fibroblasts; Abercrombie and Ambrose (1958) found that when the leading lamella of a cell touched any part of the margin of another cell its ruffling activity stopped and movement of the cell in that direction ceased. Subsequent investigations have shown that contact inhibition is a complex process involving adhesion, paralysis and contraction (Abercrombie, 1970).

A sequence of photomicrographs in which a chick heart fibroblast displays contact inhibition after colliding with another cell of the same type is shown in Figure 7.1. After colliding, the cells adhere to each other in the region of the contact. Evidence for the formation of an adhesion between cells in this manner comes from ultrastructural studies which have revealed that cytoplasmic specialisations, similar to the adhesive plaques formed between cell and substratum (see p. 152), develop where the cells touch (Heaysman and Pegrum, 1973a). In addition, when the cells subsequently separate after contact inhibition has occurred, cytoplasmic processes are often drawn out between the cells in the region of their first contact (Figure 7.1). The adhesion is accompanied by paralysis of the leading lamella; its protrusive activity

Figure 7.1: A sequence of phase contrast photomicrographs, taken at 10 minute intervals, showing contact inhibition between two chick heart fibroblasts. (R.A. Peachey, dust cover for *Cell Behaviour*, R. Bellairs, A. Curtis and G. Dunn (eds.), Cambridge University Press, 1982.)

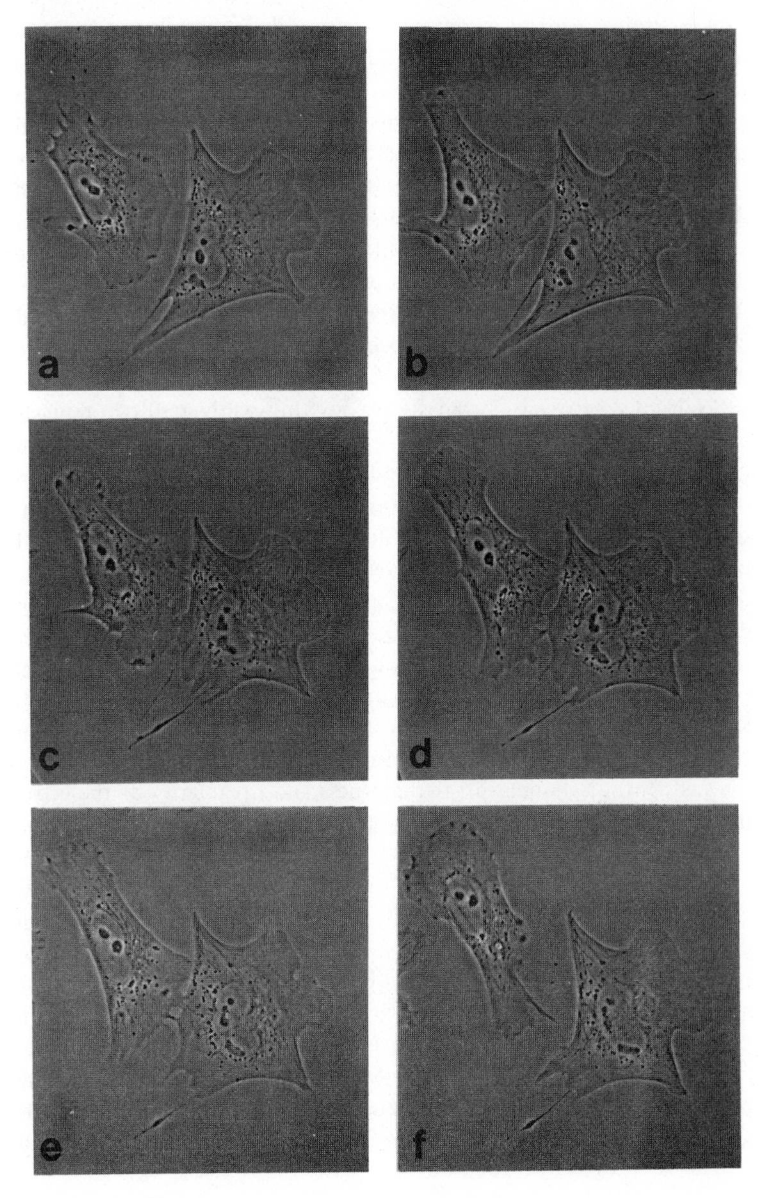

and ruffling stop (Abercrombie, 1970). In the example illustrated here the leading lamella makes contact with the other fibroblast over a broad front, and as a result paralysis is extensive (Figure 7.1). Often, however, only a part of the lamella touches another cell, and in such cases paralysis is confined to that part of the lamella and ruffling and protrusion continue unabated in adjacent regions of the lamella not involved in the contact (Trinkaus *et al.*, 1971). At about the same time as its locomotory activity is paralysed, the leading lamella contracts. This contraction is also localised but involves a considerable proportion of the lamella behind its front edge and may be strong enough to separate the cells (Abercrombie, 1970). While these events are happening, a new leading lamella is often being formed elsewhere on the cell. In Figure 7.1 the new leading lamella forms from what had previously been the 'tail' of the cell, and the direction of locomotion is therefore reversed. Alternatively, a lateral portion of the original leading lamella not in contact with the other cell may become dominant and change the cell's direction of movement accordingly (Abercrombie, 1970).

Collisions between fibroblasts do not invariably result in contact inhibition, and the frequency with which it occurs varies in different cell types and with different experimental procedures. It is therefore useful to have some method of assaying the phenomenon quantitatively; the only reliable way of doing this is to observe a series of collisions between cells and to record their outcome. It is, unfortunately, a tedious and time-consuming process to collect adequate data using this approach, and quicker but more indirect methods have therefore been developed. Since contact inhibition should prevent one cell from overlapping another, the extent to which a population exists as a monolayer can be used as an assay for the phenomenon. Fixed and stained cultures are usually examined to determine the extent to which the cells form a monolayer; the faintness and irregularity of the edges of the cells make it difficult to detect cytoplasmic overlaps, so overlaps between nuclei are usually counted. The number of nuclear overlaps is expressed as a percentage of the number expected if the nuclei had been distributed at random, thus providing an overlap index for the population of cells (Abercrombie and Heaysman, 1954; Abercrombie *et al.*, 1968). An overlap index of 100 per cent indicates that the cells are randomly arranged and any figure less than this implies that the cells are to some extent inhibited from moving over each other; the more closely the index approaches zero the more perfectly is the population organised as a monolayer.

A homotypic overlap index reflects the extent to which a population of a single type of cell is 'monolayered' and with slight modification similar methods can be applied to derive a heterotypic overlap index expressing the extent to which two different types of cell overlap each other (Abercrombie *et al.*, 1968). A low overlap index is often interpreted as evidence that the cells exhibit contact inhibition, and a high index as evidence for the opposite, but such inferences must be treated with caution since factors other than contact inhibition may lead to monolayering, while extensive overlapping of cells may not necessarily imply the absence of contact inhibition. For example, mixed cultures of chick fibroblasts and rabbit polymorphonuclear leucocytes have a low heterotypic overlap index (Armstrong and Lackie, 1975), which could be interpreted as meaning that collisions between these different cell types result in contact inhibition. However, direct observation of the cells shows that such collisions do not result in the changes characteristic of contact inhibition, protrusive activity and ruffling do not stop and there is no contraction (Armstrong and Lackie, 1975). Obviously factors other than contact inhibition must prevent these cells from overlapping each other. Similar monolayering in the absence of contact inhibition has been described in a number of other types of cell (see Heaysman, 1978, for a review). In each case contact with another cell does not inhibit the locomotory activity of the cell but overlapping does not occur, probably because the surface of one cell does not provide a suitable substratum for the movement of the other. Such monolayering without inhibition of locomotory activity has variously been termed 'type 2 contact inhibition' (Vesely and Weiss, 1973), 'contact inhibition of the second kind' (Harris, 1974), and 'contact inhibition of overlapping' (Garrod and Steinberg, 1975). But Heaysman (1978) has proposed that the term contact inhibition be retained for the contact interaction resulting in an inhibition of cellular locomotory activity, and that the monolayering of cells without such inhibition should be termed 'substratum-dependent inhibition of locomotion'. We shall adopt this nomenclature.

At the other extreme, a high overlap index need not be associated with a reduced level, or failure, of contact inhibition. If there is a sufficient gap between part of a cell and the substratum another cell may be able to move into this space without the surfaces of the two cells coming into contact. In fixed cultures the cells would seem to be overlapping, but since they have not made contact it would be wrong to assume that this was the result of the absence or failure of contact inhibition. Again, if extracellular material, such as collagen, is secreted by cells *in vitro* it may coat their surfaces and prevent effective contacts between them, and any overlapping could not be interpreted as due to the absence or failure of contact inhibition (Abercrombie, 1970).

In view of these qualifications, inferences drawn from overlap indices about the occurrence (or otherwise) of contact inhibition between cells should always be confirmed by direct observation of the effects of cell contact on the locomotory activity of the cells.

Since its original discovery in cultures of embryonic chick heart fibroblasts, contact inhibition has been shown to occur *in vitro* between normal fibroblasts derived from a wide variety of embryonic and adult tissues (see Abercrombie, 1970; Heaysman, 1978). It occurs not only when fibroblasts of the same type collide (i.e. homotypic contact inhibition) but also when fibroblasts derived from different tissues, or even from different animals, collide; for example collisions between fibroblasts from embryonic chick heart and from new-born mouse muscle result in mutual heterotypic contact inhibition (Abercrombie *et al.*, 1968). Furthermore 'self-contact inhibition' in cultures of chick heart fibroblasts has been described (Ebendal and Heath, 1977). This occurs when two lamellae from the same fibroblast make contact with each other; both lamellae are paralysed and contract as if they belonged to two different cells.

Contact Inhibition in Epithelial Cells

Although they have been studied less intensively than fibroblasts, there is some evidence that epithelial cells *in vitro* also display contact inhibition. Unlike fibroblasts, epithelial cells may move over a plane substratum either as single isolated cells or as coherent sheets (Middleton, 1973); the same cells can, depending upon circumstances, behave in either of these ways. The consequences of collisions between isolated epithelial cells have not yet been fully analysed, partly because many such cells fail to show active locomotion, and also because any interaction that occurs between these cells is obscured by the strength of the adhesions that develop between them after a collision. There is, nevertheless, reason to suppose that collisions between homologous epithelial cells result in contact inhibition. Dissociated epithelial cells from a number of different sources are strongly monolayered in culture, and there is some evidence that collisions between them result in adhesion, paralysis and contraction similar to that seen in fibroblasts (see Middleton, 1982, for a review). As yet this evidence is fragmentary, and further work needs to be done; we have no information about the results of collisions between heterologous epithelial cells.

The evidence for the occurrence of contact inhibition within and between sheets of epithelial cells is more persuasive. If contact inhibition occurs between the cells in an epithelial sheet only the cells at the free edges of the sheet would be expected to display leading lamellae since they are the only cells with part of their margin not in contact with other cells. Cells not at the edge of the sheet should only form leading lamellae if, for any reason, the continuity of the sheet is disrupted and a gap appears within it; these lamellae should then develop only where the edge of a cell forms part of the boundary of the gap. Further, if a previously contact-free region of cell margin makes contact with another epithelial cell (due, for example, to the closure of a gap in the sheet, or to a collision with another sheet), the formation of leading lamellae should be inhibited in that region. The observed behaviour of a variety of epithelia *in vitro* consistently matches these predictions.

That leading lamellae are usually present only at the edges of sheets or islands (i.e. small groups) of epithelial cells has been known since the earliest days of tissue culture (Holmes, 1914) and has frequently been confirmed since then (see Middleton, 1982, for a review). Similarly, the formation of leading lamellae by the newly exposed edges of cells in the margins of spontaneously occurring or deliberately created gaps in epithelial sheets has been described by several workers (e.g. Wilbur and Chambers, 1941; Vaughan and Trinkaus, 1966). In addition, it is well known that the formation of a leading lamella is inhibited in areas of contact between homologous epithelial sheets colliding in culture (e.g. Abercrombie and Middleton, 1968; Flaxman and Nelson, 1974). In sum, all these observations are consonant with the belief that contact inhibition occurs between the individual cells in epithelial sheets, and between sheets of homologous epithelial cells *in vitro*. There is also some rather limited evidence that collisions between heterologous epithelial sheets result in contact inhibition (Abercrombie and Middleton, 1968).

Collisions between isolated epithelial cells and fibroblasts have not yet been studied in detail, but in a number of cases fibroblasts have been seen to display contact inhibition after colliding with an epithelial sheet (Middleton, 1982). Parkinson and Edwards (1978) found that fibroblasts from the chick choroid show contact inhibition after colliding with sheets of pigmented retina epithelial cells but that there is no reciprocal response from the epithelial cells; the leading lamellae at the edge of the epithelial sheet are quite unaffected by contact with the fibroblasts and the sheet continues its forward movement without interruption. This is a clear example of non-reciprocal contact inhibition, a phenomenon first described in collisions in culture between normal chick fibroblasts and mouse MCIM sarcoma cells (Heaysman, 1970, see p. 148). It remains to be seen whether such a non-reciprocal response is common between cultured epithelial cells and fibroblasts.

The Significance of Contact Inhibition

Contact inhibition influences the social behaviour of cells in culture in several ways. As we have seen, populations of cells exhibiting contact inhibition tend to be monolayered. In addition, in cultures with a uniformly high population density contact inhibition tends to immobilise the cells because movement is inhibited in all directions by contact with other cells. When the population density is not uniform contact inhibition causes cells to move away from a densely populated area towards a sparsely populated one (see Abercrombie, 1980); cell locomotion occurs preferentially in this way because those cells which move towards the sparsely populated area will be less inhibited by contact with other cells. This effect is clearly seen in the centrifugal migration of cells from an explant of tissue *in vitro*; chick heart fibroblasts, for example, spend 72 per cent of the time moving away from the explant and only 28 per cent moving towards it (Abercrombie and Heaysman, 1966). Similar orientated movement is seen when a dense and essentially immobile monolayer of cells is 'wounded' by scraping away some cells to create a cell-free space on the substratum. The cells at the edge of such a wound start to move because they are no longer in contact with neighbouring cells on all sides, and their movement is directed towards the cell-free space because movement in any other direction is inhibited by existing contacts (Vasiliev *et al.*, 1969). This directed movement persists until the space is repopulated, when it ceases gradually as the moving cells begin to collide with other cells with increasing frequency.

All this suggests that, to some extent at least, the initiation, direction and cessation of cell locomotion in culture can be explained on the basis of contact inhibition, and it is possible that, in some situations, the same is true *in vivo*. It was pointed out in Chapter 1 that embryonic development requires the active movement of both individual cells and sheets of cells, and contact inhibition could be important in regulating and directing such movement. Postnatally, the great majority of cells are normally immobile, but many still retain the potential ability to engage in cell locomotion as is obvious when they are cultured *in vitro*; it is evident that, *in vivo*, some influence normally restrains the cells from expressing their potential for moving. Since the component cells of a tissue *in vivo* often have extensive contacts with their neighbours, contact inhibition could provide such an influence; if so, cells which are usually immobile should begin to move if they lose contact with their neighbours. There is some evidence that this happens. When a tissue *in vivo* is wounded the cells at the edge of the wound start to migrate; this locomotion is apparently triggered by wounding, is directed towards the wound, and stops when tissue continuity has been restored (Abercrombie, 1964). This characteristic pattern of behaviour *in vivo* is very similar to that seen during the filling-up of a 'wound' *in vitro*, and the explanation of the latter offered above could apply equally to the healing of wounds *in vivo*.

Unfortunately, because of the difficulties inherent in observing cells as they move inside the body it is not yet clear whether cells *in vivo* do display contact inhibition. In a few fortunate situations it has been possible to watch the behaviour of embryonic cells during morphogenesis. The embryos of the annual cyprinodont fish, *Nothobranchius korthausae*, are transparent, and the blastomeres can be observed during development. In the course of epiboly the deep blastomeres migrate as individual cells into the narrow space between the expanding enveloping layer and the periblast. When these cells collide with each other they stop moving and do not overlap; usually one or both of the cells involved in a collision change direction and separate (Van Haarlem, 1979; Lesseps, Hall and Murmane, 1979). This obviously supports the possibility that these cells display contact inhibition *in vivo*, but when they are isolated from the embryo and cultured *in vitro* they behave differently. They usually fail to spread on the substratum and instead adhere to each other, forming a spherical aggregate (Lesseps, Lapeyre and Hall, 1979), suggesting that their behaviour *in vivo* is modified to some extent by interactions with neighbouring cells in the embryo. The situation is further complicated by the fact that although the locomotion of these cells *in vivo* is apparently inhibited by contacts between them, it is not inhibited by their contacts with the cells of the enveloping layer to which they are attached during their migration (Van Haarlem, 1979).

An isolated observation has been described of a collision between fibroblasts migrating in the relatively transparent cornea of the chick; this resulted in a reaction between the colliding cells similar to contact inhibition (Bard and Hay, 1975). These cells are known to exhibit contact inhibition *in vitro* (Bard and Hay, 1975).

It has also been possible to observe directly the behaviour of cells in the epithelial sheets which migrate during the expansion of the chick embryo blastoderm and during epiboly in *Fundulus*. Leading lamellae are usually formed only by cells at the free edges of these sheets, and cells entirely within the sheets only develop lamellae if the formation of a gap in the sheet causes some of their margin to lose contact with adjacent cells; these lamellae disappear when contact is re-established as the gap is closed (Trinkaus and Lentz, 1967; Bellairs *et al.*, 1969). This is similar to the behaviour of epithelial cells in culture described above, and lends support to the possibility that contact inhibition is also a feature of the locomotion of epithelial cells *in vivo*.

It has been relatively easy to study the behaviour of cells during the healing of wounds *in vivo*, and attention has again been directed chiefly to the movements of epithelial cells. It has been known for some time that a sheet of epithelial cells migrating across a wound continues to advance until it makes contact with the sheet coming from the opposite edge. When contact is established the migration of both sheets stops, preventing their superimposition (Lash, 1955; Chiakulas and Millman, 1959). In a recent elegant experiment Radice (1980a) has observed this process in greater detail in skin wounds in *Xenopus* tadpoles. Prior to the infliction of a wound, the basal cells of the epidermis are immobile, but 5-10 seconds after the wound has been made the basal cells at its edges extend leading lamellae over the surface of the wound and begin to move across it. This movement goes on until contact is established between the leading lamellae of the sheets of epithelium advancing from opposite sides, and shortly afterwards the protrusive activity of the leading lamellae stops and cell movement ceases; hence the opposing sheets do not overlap each other (Radice, 1980a). These cells show very similar behaviour *in vitro* (Radice, 1980b), and the results of these investigations provide another piece of evidence in support of the proposition that contact inhibition occurs *in vivo* and may be important in controlling and directing cell movement in the body as well as in cultures.

Since contact inhibition can immobilise cells, it could account for the fact that, *in vivo*, most populations of cells do not normally move into an area already occupied by another population. But certain cells do show invasive behaviour; this is normally displayed by some leucocytes, and abnormally, of course, malignant cancer cells intrude into surrounding normal tissues. Thus cells which are invasive *in vivo* might be expected to reveal deficient contact inhibition *in vitro*. It is important to appreciate that when assessing whether or not one population of cells will invade another population *in vitro*, it is the outcome of heterotypic collisions that is significant. Either of the two chosen types of cell may or may not exhibit a high degree of homotypic contact inhibition, but the extent to which one type may invade the other depends mainly on the extent to which they display heterotypic contact inhibition.

Polymorphonuclear leucocytes (PMNs) are an example of a type of cell which normally invades tissues *in vivo* to carry out their function of phagocytosis and digestion of bacteria or other 'foreign' materials. They are able to migrate through capillary walls and become actively motile in the tissue spaces outside the blood vessels. This invasive behaviour is retained in organ culture and populations of PMNs will invade aggregates of normal fibroblasts (Armstrong and Lackie, 1975). If there is a correlation between invasiveness and a deficiency in heterotypic contact inhibition, collisions in culture between fibroblasts and PMNs should show an absence of contact inhibition or at least a significant reduction in its extent. This has been confirmed; chick PMNs invade outgrowths of normal fibroblasts from explants of chick embryo heart *in vitro* and the heterotypic overlap index in such cultures is high (Oldfield, 1963). Careful analysis of cine films of mixed cultures of rabbit PMNs and chick fibroblasts confirms that collisions between the two cell types do not result in the changes characteristic of contact inhibition (Armstrong and Lackie, 1975). Confusingly, with this particular combination of cells the heterotypic overlap index does suggest that overlapping of the two cell types is inhibited, but detailed analysis has revealed that this is the result of substratum-dependent inhibition of locomotion (see p. 141) and is not caused by heterotypic contact inhibition (Armstrong and Lackie, 1975). In the case of polymorphonuclear leucocytes, therefore, their invasiveness *in vivo* correlates with their deficient heterotypic contact inhibition *in vitro*.

Malignant tumour cells *in vivo* infiltrate surrounding normal tissues, resulting in local spread of the cancer and leading to metastasis (i.e. the development of secondary tumours at distant sites). Here, too, there seems to be an association between invasiveness *in vivo* and impaired heterotypic contact inhibition *in vitro*; cells derived from a number of sarcomas which are known to be invasive *in vivo* have been shown to exhibit deficient heterotypic contact inhibition *in vitro*. In simple terms these cells may be thought of as malignant fibroblasts, since sarcomas are defined as malignant tumours arising from non-epithelial mesodermal tissues. In culture, populations of normal fibroblasts migrating from adjacent explants of chick embryo heart and neonatal mouse muscle do not invade each other; the heterotypic overlap index is low and analysis of time-lapse films shows that collisions between cells of the two types result in mutual heterotypic contact inhibition (Abercrombie *et al.*, 1957; Stephenson, 1982); the locomotory activity of each type of cell is inhibited by contact with the other type. But when explants of mouse sarcoma S180 are cultured close to explants of chick embryo heart, under the same conditions, the fibroblasts migrating from the two explants *do* invade each other. The heterotypic overlap index of such cultures indicates that the number of overlaps between the two types of cell is not significantly different from that which would occur if the cells were distributed at random (Abercrombie *et al.*, 1957). In this case time-lapse films show that collisions between cells of the two kinds do not lead to any of the typical effects of contact inhibition in either cell (Abercrombie and Ambrose, 1958). In other words, the two populations invade each other in culture because both fail to display heterotypic contact inhibition. Similar *in vitro* behaviour is shown by cells derived from four other mouse sarcomas (311, BAS56, FS9 and MCIM), but in none of these does invasion result merely from a similar reciprocal absence of contact inhibition. In all four cases, the heterotypic overlap index between the tumour cells and normal chick heart fibroblasts is greater than that found between normal mouse fibroblasts and chick heart fibroblasts, but less than that expected if the two populations were distributed at random, so it would appear that heterotypic contact inhibition is not totally absent between these tumour cells and normal fibroblasts (Abercrombie and Heaysman, 1976; Abercrombie, 1979). Further investigation has shown that collisions between normal chick fibroblasts and cells from BAS56 and 311 sarcomas result in some degree of mutual heterotypic contact inhibition, but this is not sufficient to prevent normal and tumour cells invading each other (see Abercrombie, 1979, for a review). In these instances, then, invasion results from a reduced level, rather than a complete absence, of heterotypic contact inhibition.

Collisions between normal chick heart fibroblasts and either FS9 or MCIM sarcoma cells result in non-reciprocal contact inhibition. Cine films reveal that the locomotory activity of the normal cells is inhibited by contact with either of the types of tumour cell, whereas the behaviour of the sarcoma cells is unchanged by contact with the normal cells (Heaysman, 1970; Abercrombie, 1979). Here the cancer cells show an absence of contact inhibition and invade the normal cells, whereas the latter are still subject to inhibition and cannot invade the cancer cells.

Because of the results obtained with sarcoma cells, malignant cells in culture are often described as lacking contact inhibition, but this is a misunderstanding. We have seen that the invasive behaviour of sarcoma cells in culture appears to result from a deficiency in heterotypic contact inhibition, but it does not necessarily follow that these cells also exhibit deficient homotypic contact inhibition. It has been pointed out (see p. 144) that the orientated migration of normal cells from an explant can be accounted for by homotypic contact inhibition; malignant cells migrate from explants in much the same way as normal cells and are therefore presumably subject to the same mechanism. A reduced level of homotypic contact inhibition between malignant cells would be expected to reduce rather than to increase their invasive potential, because their locomotion would then tend to become random rather than being orientated in a particular direction. Measurements of homotypic contact inhibition in cultures of malignant cells confirm that they display this type of inhibition, but to a more variable extent than normal cells (Abercrombie, 1979; Projan and Tanneberger, 1973).

Epithelial cells are the source of some of the commonest malignancies in man (carcinomas). It is obviously desirable to establish whether the invasive behaviour of epithelial cancers may also be accounted for by deficiencies in their heterotypic contact inhibition but, unfortunately, contact interactions between carcinoma cells and normal cells *in vitro* have not been studied adequately, and we cannot as yet answer this question.

To summarise, there is no doubt that contact inhibition of locomotion plays a part in regulating the movement of cells in culture, and there is at least preliminary evidence that the same may be true *in vivo*. It has been proposed that:

(a) Within a tissue, mutual contact inhibition prevents the majority of cells from moving.
(b) If the continuity of cell-cell contacts within a tissue is breached (e.g. by a wound), the cells bordering the breach will be released from contact inhibition. Cell locomotion will be initiated, its direction will be determined by contact inhibition, and it will cease when continuity of the tissue is restored.
(c) The movements of some normal cells, and of malignant cells, are not so markedly inhibited by contact with other cells; this deficiency in heterotypic contact inhibition enables these cells to invade surrounding tissues. Since contact inhibition results from contacts between cell surfaces, a corollary of this is that there is something different about the surfaces of cells which are able to invade tissues *in vivo*. In the case of malignant cells this further implies that the

transformation from the normal to the malignant state involves a change in the properties of the cell surface. This is consistent with the large body of evidence that the surfaces of many varieties of tumour cells are altered in several ways as compared with normal cells (e.g. Nicolson and Poste, 1976).

The Mechanism of Contact Inhibition

We can at present do no more than speculate about the nature of the mechanisms which underly contact inhibition. Any attempt to explain the phenomenon must account for the adhesion, paralysis and contraction which characterise it, and must deal with both non-reciprocal and self-contact inhibition.

A possible explanation is that one cell acts as a physical obstacle across which another cell cannot move. Leading lamellae may be unable to bend sufficiently to surmount another cell, or, if bent sufficiently, may then be unable to maintain locomotory activity (Abercrombie, 1970). This seems to be ruled out, however, by the fact that if colliding mouse fibroblasts (which are known to display homotypic contact inhibition) are fixed with glutaraldehyde, other living cells of the same type will move freely over their surfaces without any inhibition (Vesely and Weiss, 1973).

Another tentative explanation is that the surface of one cell may not be sufficiently adhesive to other cells to allow them to move over it. Alternatively, the surface of a cell might be sufficiently adhesive, but less so than the culture substratum; this could prevent a cell from detaching itself from the substratum and moving over the less adhesive surface of another cell (Abercrombie, 1970; Martz and Steinberg, 1973). This explanation for contact inhibition is supported by the observation that when cells are cultured on a substratum of varying adhesiveness they move preferentially away from regions to which they are less adhesive and congregate in regions to which they are more adhesive; the converse behaviour is not normally seen (Carter, 1965; Harris, 1973). If this reaction to variations in adhesiveness is a factor in contact inhibition, cells should display contact inhibition when they encounter the boundary between an area where the substratum is adhesive and an adjacent area where the substratum has been treated to render it less or non-adhesive. Such boundaries can be created by covering part of the culture substratum with, for example, a film of phospholipid, or with agarose (Ivanova and Margolis, 1973; Heaysman and Pegrum, 1982). However, cells which normally exhibit homotypic contact inhibition are not contact inhibited when they meet such a boundary; their forward movement stops, but the characteristic triad of adhesion, paralysis and contraction which typifies contact inhibition is not seen. Far from being paralysed, indeed, the cells' leading lamellae continue their protrusive activity unabated and vigorous ruffling often occurs as the cells vainly try to continue their locomotion (Heaysman and Pegrum, 1982).

The cessation of forward movement without true contact inhibition seen when some types of cell collide in culture (see p. 141) may similarly result from the failure of the surface of the cells to provide sufficient adhesion to permit other cells to move across them. Heaysman (1978) has suggested that such cessation of forward locomotion without an associated inhibition of locomotory activity, whether induced by contact with inert non-adhesive substrata, or by contact with other cells, should be termed 'substratum dependent inhibition of locomotion'. It is clear that this is a separate phenomenon from contact inhibition, and the latter cannot be explained in terms of differences in the ability of cells to adhere to culture substrata or to the surfaces of other cells.

This view receives further support from the observation that contact inhibition between fibroblasts often occurs as one cell attempts to move between another cell and the substratum; the moving cell then shows contact inhibition without attempting to leave the substratum, so that the adhesiveness of the surface of the other cell must be irrelevant here (Abercrombie, 1979). In a few cases the monolayering of cell populations can be attributed to substratum-dependent inhibition of locomotion (see p. 147), but this is not the explanation of monolayering where this results from contact inhibition.

Another hypothesis is that some kind of signal is exchanged between colliding cells which triggers the changes characteristic of contact inhibition. The signal might be in the form of a substance capable of inhibiting locomotory activity, which might be exchanged through permeable connections between the two cells (Abercrombie, 1970). There is certainly evidence that molecules can be exchanged between cells, both *in vivo* and *in vitro*, via specialised cell contacts known as gap junctions (reviewed by Loewenstein, 1979; Hertzberg *et al.*, 1981), but there is none to suggest that a similar exchange of molecules is involved in contact inhibition. Indeed, it is difficult to conceive how a substance capable of inhibiting locomotory activity in a recipient cell could avoid inhibiting the activity of the cell producing it. This would presumably require the substance to be synthesised and stored in an inactive form, and to be activated during the passage from one cell to the other. In addition, non-reciprocal contact inhibition, and the ability of some cells to exhibit homotypic, but not heterotypic, inhibition, are difficult to explain in terms of this 'signal' hypothesis. In spite of this it remains an attractive possibility, particularly since it is known that gap junctions can form quite quickly between some cells after they have collided in culture, and also because some malignant cells have been shown to have a reduced ability to exchange molecules with neighbouring cells (see Loewenstein, 1979).

Figure 7.2: Electron micrograph of a vertical longitudinal section through the overlapping leading lamellae of two chick embryo fibroblasts 20 seconds after their initial contact. Specialisations have developed in the cortical cytoplasm of both cells. (From Heaysman and Pegrum, 1973a.)

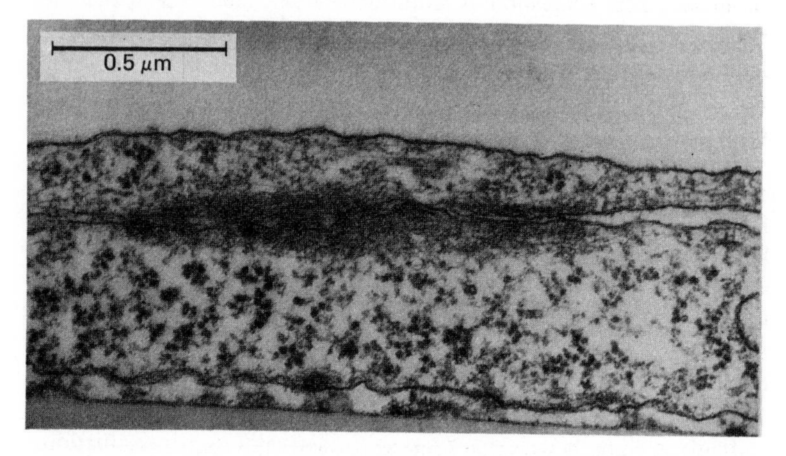

In order to learn more about such contacts, electron microscopy has been used to study cells before and after they have collided with other cells in culture and it shows that specialisations develop in the cortical cytoplasm in the region of the contact when embryonic chick heart fibroblasts exhibit homotypic contact inhibition (Heaysman and Pegrum, 1973a). These specialisations occur in both the cells involved in the collision within 20 seconds of contact being made (Figure 7.2), and appear very similar to the plaques or focal contacts formed between these cells and their substratum (see Chapter 6, p. 113). Within 60 seconds of the cells making contact microfilaments running parallel with the long axis of the cell are found associated with these specialisations and they become more distinct by 2 minutes after contact. As the contraction (which forms an essential part of contact inhibition) separates the cells the cortical specialisations and their microfilaments disappear (Heaysman and Pegrum, 1973a).

When they form between colliding fibroblasts, these specialisations are apparently associated with the cessation of locomotory activity which is a feature of contact inhibition; but very similar specialisations (plaques) between a fibroblast and its substratum are associated with active locomotion of the cell. It is not known why such similar structures should be associated with such diverse patterns of behaviour. Heaysman and Pegrum (1982) have speculated that the microfilaments associated with these specialisations might be capable of generating tension. If associated with cell-substratum contacts, such tension might contribute to the forward movement of the cell, but if caused by cell-cell contacts it might lead to contraction and thus cause the contact to break. This suggestion is consistent with Abercrombie's (1980) that contact with another cell causes contact inhibition because it induces a contraction in the leading lamella which locally suppresses the forward flow of material in the cytoplasm which is required for protrusion (see Chapter 6, p. 125).

If the formation of specialisations within the cortical cytoplasm is causally related to contact inhibition, such specialisations should not be formed when cells that do not exhibit contact inhibition collide. This has been shown to be the case in one instance. As we have seen, sarcoma S180 cells and chick embryo heart fibroblasts show a mutual lack of heterotypic contact inhibition. Electron microscopy reveals that contacts between these cells do not result in the formation of cytoplasmic specialisations in either of the cells (Heaysman and Pegrum, 1973b).

It is too early to conclude that the formation of such specialisations is invariably associated with contact inhibition of locomotion, or that failure to form them guarantees the absence of contact inhibition. It would be instructive to examine collisions that result in non-reciprocal contact inhibition to see whether the formation of cytoplasmic specialisations is also one-sided.

Contact-induced Spreading of Cells in Culture

The behaviour of epithelial cells in culture differs in some ways from that of fibroblasts. When dissociated epithelial cells are cultured on a plane substratum they characteristically form islands or sheets of closely adherent cells in contrast to the loose networks of cells formed by fibroblasts. Time-lapse films show that this pattern of behaviour is due to the fact that contacts formed between epithelial cells are more stable than those between fibroblasts. Since epithelial cells tend to remain in contact with each other after they have collided (unlike fibroblasts, which tend to move apart), there is, with increasing time *in vitro*, a gradual reduction in the number of single isolated epithelial cells and a concomitant increase in the number of cells incorporated into islands or sheets (Middleton and Pegrum, 1976).

Several different kinds of epithelial cells have been seen to behave quite differently when they are isolated as compared with their behaviour when they form part of an island or a sheet. Isolated epithelial cells often do not assume the polarised morphology so often shown by isolated fibroblasts under similar conditions. They attach firmly to the substratum but usually fail to flatten or spread extensively. They normally lack a well defined leading lamella but they display vigorous surface activity in the form of blebs, filopodia, or short ruffling lamellae. Consequently they do not undergo extensive locomotion but tend to oscillate over a limited area of the substratum (Trinkaus, 1963; Middleton, 1977; Garrod and Steinberg, 1975). In contrast to this, epithelial cells which have been incorporated into islands are usually extensively spread on the substratum; they do not bleb and, if positioned at the edge of the island, they often develop leading lamellae (Trinkaus, 1963; Middleton, 1977; Di Pasquale, 1976). This difference in behaviour between isolated cells and cells in contact with neighbours seemed likely to be due to some form of contact interaction, and an investigation of the behaviour of chick embryo pigmented retina epithelial cells *in vitro* has confirmed that this is the case. It was found that when isolated cells of this type made contact with an island of similar cells they underwent a striking change in their morphology and behaviour, becoming indistinguishable from the rest of the cells in the island (Middleton, 1977). A sequence of photomicrographs illustrating this is shown in Figure 7.3.

Prior to making contact with the island the single isolated cell is poorly spread, its nucleus is obscured by pigment granules, it lacks a leading lamella and it is blebbing vigorously (Figure 7.3a). Within an hour of establishing contact with the island of cells its blebbing has decreased and it has begun to spread more extensively on the substratum; as a result its nucleus is less obscured by the pigment granules (Figure 7.3b). Two hours after joining the island the cell has spread to a much greater extent and the nucleus is clearly visible, blebbing has stopped and a leading lamella is evident in that part of its margin which is free of contact with other cells (Figure 7.3c). Three hours after

Figure 7.3: A sequence of phase contrast photomicrographs, taken at 1 hour intervals, of living chick embryo pigmented retinal epithelial cells. The cell marked by the arrowhead shows contact induced spreading. (From Middleton, 1977.)

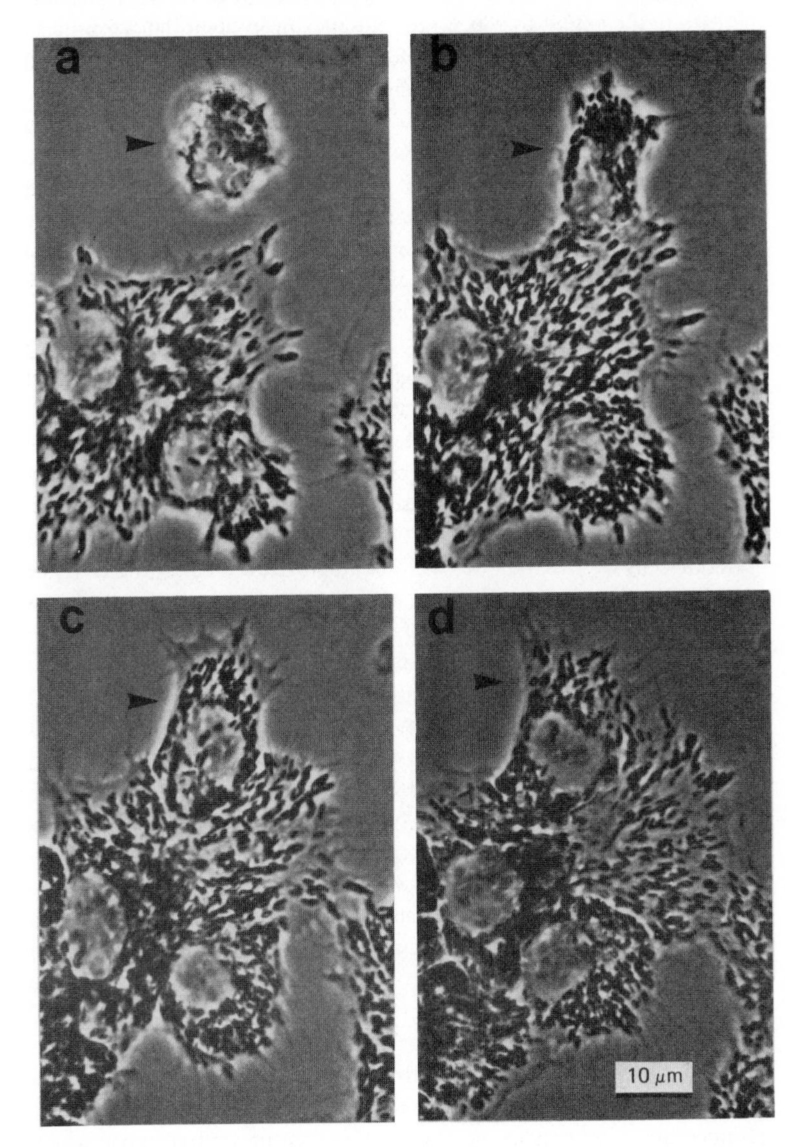

contact the cell is indistinguishable from its neighbours in the island (Figure 7.3d).

This alteration in the behaviour of initially isolated cells seems to be a direct response to contact with an island and, since it results in an increase in the extent to which the cells are spread upon the substratum it has been named 'contact-induced spreading' (Middleton, 1977). Contact-induced spreading occurs not only when an isolated cell makes contact with an island of cells but also when two isolated cells make contact with each other (Middleton, 1977), and this provides the basis for a convenient quantitative assay of the phenomenon. Within a single culture the area of substratum (spread area) occupied by isolated cells and the area occupied by islands consisting of two cells can be measured and compared; if the cells exhibit contact-induced spreading the spread area of each of the cells in a two cell island should be greater than the spread area of an isolated cell. By this means it has been shown that after 24 hours, in cultures of pigmented retina epithelial cells, each cell in a two cell island has an average spread area that is approximately 2.2 times that of isolated cells (Middleton, 1977). A similar investigation has shown that in cultures of chick embryo corneal epithelium the cells in islands of two cells have significantly larger spread areas than isolated cells, and direct observations have confirmed that collisions between cells of this type result in contact-induced spreading (Brown and Middleton, 1981).

Other studies suggest that epithelial cells from a variety of sources display contact-induced spreading in culture, but since it has not been described in cultures of fibroblasts it seems at present that the phenomenon may be restricted to cells of an epithelial nature (see Middleton, 1982, for a review).

So far, contact-induced spreading has only been shown to occur when epithelial cells of the same type collide in culture, and it is not yet known whether it applies to collisions between epithelial cells of different types or to collisions between epithelial cells and fibroblasts; nor is it known if it occurs when epithelial cells collide *in vivo*.

Contact-induced spreading is completely reversible, and when, as happens infrequently, cells break away from an island of epithelium or from a sheet, they quickly revert to the behaviour characteristic of isolated cells; their leading lamellae disappear, their spread area diminishes, and they start to bleb (Di Pasquale, 1975; Garrod and Steinberg, 1975; Middleton, 1977). However, it is not an all-or-none phenomenon; in cultures of corneal epithelium and in cultures of pigmented retina epithelium from chick embryos a small proportion of isolated cells lack blebs and are extensively spread, with spread areas similar to those of cells in islands (Middleton, 1977; Brown and Middleton, 1981). In fact, time-lapse films have shown that isolated cells of these types can vary the extent to which they are spread (in the absence of contacts with other cells). An individual isolated cell may sometimes exhibit a well spread morphology, while at others it may be only poorly spread; in general, though, a given cell will be in a poorly spread state for most of the time (Middleton, 1982). Thus contact with another cell is not an absolute prerequisite for the spreading of these epithelial cells. It appears that the effect of cell contact is to induce and stabilise a well spread morphology that can be more transiently displayed by isolated cells (Brown and Middleton, 1981).

The mechanisms which operate during contact-induced spreading are unknown, but they result in changes in several aspects of cell behaviour. The affected cell makes contact with, and adheres to, another cell and spreads more extensively on the substratum, so changes in both cell-cell and cell-substratum adhesion must be involved. The cell also becomes polarised, in the sense that generalised surface activity (blebbing) ceases and a leading lamella develops from a localised region of the cell margin. It may be that this polarisation precedes the more extensive spreading of the cell and may direct its locomotory activity away from the contact it has formed (Middleton, 1977; Brown and Middleton, 1981). Perhaps the locomotory force that the cell exerts against this contact results in the spreading of the cell (Brown and Middleton, 1981). Since contact-induced spreading seems to be a direct response to cell contact it has been suggested that it may share mechanisms in common with contact inhibition (Middleton, 1977; 1982). However, the latter, although capable of directing the movement of the cell as a whole, is known to be the result of a localised change in the locomotory activity of the cell margin induced by cell-cell contact; as yet it has not been demonstrated that contact-induced spreading results from a similar localised change.

Conclusions

Both contact inhibition and contact-induced spreading can significantly affect cell behaviour in culture, and it is possible that these are only two examples of a more general class of cell contact dependent phenomena which could play an important part in regulating cell behaviour. It is likely that contact interactions also influence cell behaviour *in vivo*. A growing body of evidence points to the possibility that the locomotion of normal cells *in vivo* may be regulated by contact inhibition, and there may well be a correlation between invasive behaviour *in vivo* and deficient contact inhibition. As yet there is no evidence to indicate that contact-induced spreading also occurs between epithelial cells *in vivo*, but, potentially, it could provide a mechanism for polarising, and thus directing, their locomotion in the intact body.

The mechanisms through which cell contacts modify the locomotory activity of cells are largely unknown, and their identification must probably await a fuller knowledge of how cells generate and co-ordinate their locomotion. This knowledge may also be fundamental to a better understanding of important processes such as morphogenesis, the healing of wounds, the development and maintenance of immunity, and the invasive activity of malignant cells.

References

M. Abercrombie (1964) 'Behaviour of Cells Towards One Another', *Adv. Biol. Skin*, vol. 5, p. 95
— (1970) 'Contact Inhibition in Tissue Culture', *In Vitro*, vol. 6, p. 128
— (1979) 'Contact Inhibition and Malignancy', *Nature*, vol. 281, p. 259
— (1980) 'The Crawling Movement of Metazoan Cells', *Proc. Roy. Soc. B*, vol. 207, p. 129
— and E.J. Ambrose (1958) 'Interference Microscope Studies of Cell Contacts in Tissue Culture', *Exp. Cell Res*., vol. 15, p. 332
— and J.E.M. Heaysman (1953) 'Observations on the Social Behaviour of Cells in Tissue Culture. I. Speed of Movement of Chick Heart Fibroblasts in Relation to Their Mutual Contacts', *Exp. Cell Res*., vol. 5, p. 111
— and J.E.M. Heaysman (1954) 'Observations on the Social Behaviour of Cells in Tissue Culture. II. "Monolayering" of Fibroblasts', *Exp. Cell Res*., vol. 6, p. 293
— and J.E.M. Heaysman (1966) 'The Directional Movement of Fibroblasts Emigrating from Cultured Explants', *Ann. Med. exp. Biol. Fenn*., vol. 44, p. 161
— and J.E.M. Heaysman (1976) 'Invasive Behaviour Between Sarcoma and Fibroblast Populations in Cell Culture', *J. Nat. Cancer Inst*., vol. 56, p. 561
—, J.E.M. Heaysman and H.M. Karthauser (1957) 'Social Behaviour of Cells in Tissue Culture. III. Mutual Influence of Sarcoma Cells and Fibroblasts', *Exp. Cell Res*., vol. 13, p. 276
—, D.M. Lamont and E.M. Stephenson (1968) 'The Monolayering in Tissue Culture of Fibroblasts from Different Sources', *Proc. Roy. Soc. B*., vol. 170, p. 349
— and C.A. Middleton (1968) 'Epithelial-Mesenchymal Interactions Affecting Locomotion of Cells in Culture', in R. Fleischmajer and R.E. Billingham (eds.), *Epithelial-Mesenchymal Interactions* (Williams and Wilkins, Baltimore), pp. 56-63
P.B. Armstrong and J.M. Lackie (1975) 'Studies on Intercellular Invasion *In Vitro*

Using Rabbit Peritoneal Neutrophil Granulocytes (PMNs). I. Role of Contact Inhibition of Locomotion', *J. Cell Biol.*, vol. 65, p. 439

J.B.L. Bard and E.D. Hay (1975) 'The Behaviour of Fibroblasts from the Developing Avian Cornea', *J. Cell Biol.*, vol. 67, p. 400

R. Bellairs, A. Boyde and J.E.M. Heaysman (1969) 'The Relationship Between the Edge of the Chick Blastoderm and the Vitelline Membrane', *Wilhelm Roux Arch. EntwMech. org.*, vol. 163, p. 13

R.M. Brown and C.A. Middleton (1981) 'Contact-Induced Spreading in Cultures of Corneal Epithelial Cells', *J. Cell Sci.*, vol. 51, p. 143

S.B. Carter (1965) 'Principles of Cell Motility: The Direction of Cell Movement and Cancer Invasion', *Nature*, vol. 208, p. 1183

J.J. Chiakulas and M. Millman (1959) 'Regeneration of the Urodele Gall Bladder Following Partial Excision', *Anat. Rec.*, vol. 133, p. 129

A. Di Pasquale (1975) 'Locomotory Activity of Epithelial Cells in Culture', *Exp. Cell Res.*, vol. 94, p. 191

T. Ebendal and J.P. Heath (1977) 'Self-Contact Inhibition of Movement in Cultured Chick Heart Fibroblasts', *Exp. Cell Res.*, vol. 110, p. 469

B.A. Flaxman and B.K. Nelson (1974) 'Ultrastructural Studies of the Early Junctional Zone Formed by Keratinocytes Showing Contact Inhibition of Movement *In Vitro*', *J. Invest. Dermatol.*, vol. 63, p. 326

D.R. Garrod and M.S. Steinberg (1975) 'Cell Locomotion Within a Contact Inhibited Monolayer of Chick Embryonic Liver Parenchyma Cells', *J. Cell Sci.*, vol. 18, p. 405

A. Harris (1973) 'Behaviour of Cultured Cells on Substrata of Variable Adhesiveness', *Exp. Cell Res.*, vol. 77, p. 285

— (1974) 'Contact Inhibition of Cell Locomotion', in R.P. Cox (ed.), *Cell Communication* (John Wiley and Sons, New York), pp. 147-85

J.E.M. Heaysman (1970) 'Non-Reciprocal Contact Inhibition', *Experientia*, vol. 26, p. 1344

— (1978) 'Contact Inhibition of Locomotion: A Reappraisal', *Int. Rev. Cytol.*, vol. 55, p. 49

— and S.M. Pegrum (1973a) 'Early Contacts Between Fibroblasts: An Ultrastructural Study', *Exp. Cell Res.*, vol. 78, p. 71

— and S.M. Pegrum (1973b) 'Early Contacts Between Normal Fibroblasts and Mouse Sarcoma Cells: An Ultrastructural Study', *Exp. Cell Res.*, vol. 78, p. 479

— and S.M. Pegrum (1982) 'Early Cell Contacts in Culture', in R. Bellairs, A. Curtis and G. Dunn (eds.), *Cell Behaviour* (Cambridge University Press), pp. 49-76

E.L. Hertzberg, T.S. Lawrence and N.B. Gilula (1981) 'Gap Junctional Communication', *Annual Rev. Physiol.*, vol. 43, p. 479

S.J. Holmes (1914) 'The Behaviour of the Epidermis of Amphibians When Cultivated Outside the Body', *J. exp. Zool.*, vol. 17, p. 281

O.Y. Ivanova and L.B. Margolis (1973) 'The Use of Phospholipid Film for Shaping Cell Cultures', *Nature*, vol. 242, p. 200

J.W. Lash (1955) 'Studies on Wound Closure in Urodeles', *J. exp. Zool.*, vol. 128, p. 13

R.J. Lesseps, M. Hall and M.B. Murmane (1979) 'Contact Inhibition of Cell Movement in Living Embryos of an Annual Fish *Nothobranchius korthausae*: Its Role in the Switch from Persistent to Random Cell Movement', *J. exp. Zool.*, vol. 207, p. 459

—, M.V. Lapeyre and M.V. Hall (1979) 'Tissue Culture Evidence on the Control of the Switch from Contact Inhibition of Cell Movement to Overlapping Behaviour in Annual Fish Embryos of *Nothobranchius korthausae*', *J. exp.*

Zool., vol. 210, p. 521
W.R. Loewenstein (1979) 'Junctional Intercellular Communication and the Control of Growth', *Biochim. Biophys. Acta*, vol. 560, p. 1
E. Martz and M.S. Steinberg (1973) 'Contact Inhibition of What? An Analytical Review', *J. Cell Physiol.*, vol. 81, p. 25
C.A. Middleton (1973) 'The Control of Epithelial Cell Locomotion in Tissue Culture', in *Locomotion of Tissue Cells* (Ciba Foundation Symposium 14, Elsevier, Amsterdam), pp. 251-70
— (1977) 'The Effects of Cell-Cell Contact on the Spreading of Pigmented Retina Epithelial Cells in Culture', *Exp. Cell Res.*, vol. 109, p. 349
— (1982) 'Cell Contacts and the Locomotion of Epithelial Cells', in R. Bellairs, A. Curtis and G. Dunn (eds.), *Cell Behaviour* (Cambridge University Press), pp. 159-82
— and S.M. Pegrum (1976) 'Contacts Between Pigmented Retina Epithelial Cells in Culture', *J. Cell Sci.*, vol. 22, p. 371
G.L. Nicolson and G. Poste (1976) 'The Cancer Cell: Dynamic Aspects and Modifications in Cell-Surface Organisation', *New Engl. J. Med.*, vol. 295, Pt. 1, p. 197; Pt. 2, p. 253
F.E. Oldfield (1963) 'Orientation Behaviour of Chick Leucocytes in Tissue Culture and Their Interaction with Fibroblasts', *Exp. Cell Res.*, vol. 30, p. 125
E.K. Parkinson and J.G. Edwards (1978) 'Non-Reciprocal Contact Inhibition of Locomotion of Chick Embryonic Choroid Fibroblasts by Pigmented Retina Epithelial Cells', *J. Cell Sci.*, vol. 33, p. 103
A. Projan and S. Tanneberger (1973) 'Some Findings on Movement and Contact of Human Normal and Tumour Cells *In Vitro*', *Eur. J. Cancer*, vol. 9, p. 703
G.P. Radice (1980a) 'The Spreading of Epithelial Cells During Wound Closure in *Xenopus* Larvae', *Dev. Biol.*, vol. 76, p. 26
G.P. Radice (1980b) 'Locomotion and Cell-Substratum Contacts of *Xenopus* Epidermal Cells *In Vitro* and *In Situ*', *J. Cell Sci.*, vol. 44, p. 201
E.M. Stephenson (1982) 'Locomotory Invasion by Tumour Cells in Explant Confrontations', in R. Bellairs, A. Curtis and G. Dunn (eds.), *Cell Behaviour* (Cambridge University Press), pp. 499-527
M.G.P. Stoker and H. Rubin (1967) 'Density Dependent Inhibition of Cell Growth in Culture', *Nature*, vol. 215, p. 171
J.P. Trinkaus (1963) 'The Cellular Basis of *Fundulus* Epiboly. Adhesivity of Blastula and Gastrula Cells in Culture', *Dev. Biol.*, vol. 7, p. 513
—, T. Betchaku and L.S. Krulikowsky (1971) 'Local Inhibition of Ruffling During Contact Inhibition of Cell Movement', *Exp. Cell Res.*, vol. 64, p. 291
— and T.L. Lentz (1967) 'Surface Specialisations of *Fundulus* Cells and Their Relation to Cell Movements During Gastrulation', *J. Cell Biol.*, vol. 32, p. 139
R. Van Haarlem (1979) 'Contact Inhibition of Overlapping: One of the Factors Involved in Deep Cell Epiboly of *Nothobranchius korthausae*', *Dev. Biol.*, vol. 70, p. 171
J.M. Vasiliev, I.M. Gelfand, L.V. Domnina and R.I. Rappoport (1969) 'Wound Healing Processes in Cell Cultures', *Exp. Cell Res.*, vol. 54, p. 83
R.B. Vaughan and J.P. Trinkaus (1966) 'Movements of Epithelial Cell Sheets *In Vitro*', *J. Cell Sci.*, vol. 1, p. 407
P. Vesely and A. Weiss (1973) 'Cell Locomotion and Contact Inhibition of Normal and Neoplastic Rat Cells', *Int. J. Cancer*, vol. 11, p. 64
K.M. Wilbur and R. Chambers (1941) 'Cell Movements in the Healing of Microwounds *In Vitro*', *J. exp. Zool.*, vol. 91, p. 287

INDEX

It is important to increase our knowledge of cell locomotion because of its central role in such vital processes as embryonic development and immunity, and its significance in cancer. An attempt is made in this book to describe the essential features of tissue culture techniques and to discuss the application of the light microscope and the electron microscope to the study of cell movement.

Attention is also devoted to cytoplasmic filaments. The structure and cytoplasmic distribution of microfilaments are described, with an account of the proteins with which they are thought to interact. Intermediate filaments are also dealt with. Similar treatment is given to microtubules, the drugs which act on them, and their assembly within the cell. The information obtained from this study is applied to the movements seen in cultured fibroblasts, and several hypotheses which have been proposed to explain these movements are considered critically. Finally, the 'social behaviour' of cells in culture is described, with particular reference to the factors which influence the movements of cultured cells, especially contact inhibition and contact induced spreading.